D0397394

1st 20ª
10ᵘ

GONE WHALING

A Search for Orcas in Northwest Waters

DOUGLAS HAND

Simon & Schuster

New York   London   Toronto

Sydney   Tokyo   Singapore

SIMON & SCHUSTER
Rockefeller Center
1230 Avenue of the Americas
New York, New York 10020

Designed by Hyun Joo Kim

Manufactured in the United States of America

1   3   5   7   9   10   8   6   4   2

Library of Congress Cataloging-in-Publication Data

Hand, Douglas, date.
Gone whaling: a search for orcas in Northwest waters /Douglas Hand
p.   cm.
1. Killer whale—Northwest, Pacific. I. Title.
QL737.C432H365   1994
599.5'3—dc20   94-4195
                                                              CIP

ISBN: 0-671-76840-9

To Marjorie and Walter

Contents

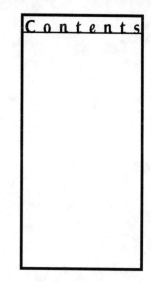

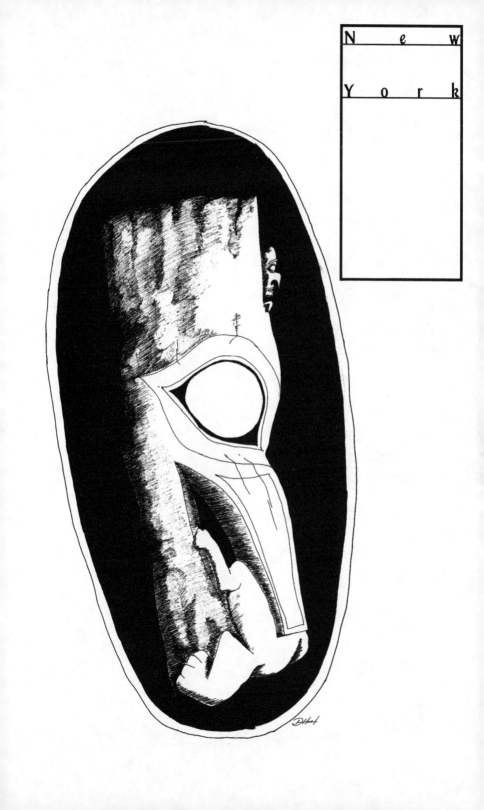

N e w

Y o r k

NEW YORK

I live in New York City now and I like it here, but I have to tell you this story.

When I arrived in New York, I decided that, to celebrate, I would go to the top of Rockefeller Center during the first snowstorm, order a drink—Grand Marnier, I thought—sit back and watch the storm swirl around me. And, in December, when that first storm blew in from the west, dumping six inches of snow in just a few hours, I pulled on my army jacket and stepped out into the wet, brittle night air. Cabs were moving at half speed, impatiently, their back ends fishtailing and their horns bleating. I took the Broadway Local train to Fiftieth Street and walked across town into the wind. Snowflakes stuck in my beard. As I rose in the elevator to the sixty-sixth floor, I imagined what awaited in precise detail: the dark, nearly empty lounge; the warmth of the chair; the burning of the liquor on my tongue and the lights of the city below, as if the stars had fallen beneath my feet. When I reached the top floor, I dusted a few remaining snowflakes from my jacket and stepped to the door. The maitre d' stopped me. He said, "I'm sorry, sir, but we can't allow anyone in without a coat and tie."

"You're kidding," I said.

"No exceptions," he said, shaking his head.

"I won't even take off my coat," I said, politely. "Nobody will ever know."

He slowly looked me up and down and, smiling, said, "Nope."

I smiled, too. I left. I never returned.

Rockefeller Center wasn't my only romantic notion. I had also dreamed of living in a garret, but the reality was cold, since my windows were rusted open. The very night I went out to sip cognac, snow was blowing in my window, which opened onto an airshaft and allowed the nightly arguments of the couple next door to drift in with the weather. One night, the woman, hands on hips, her face carved in teak, reduced a lifetime to a sentence with: "You ain't my true love, because my true love wouldn't be a motherfucker like you."

Yet, if there was any truth in my romance, I knew it was

this: At that point in my life, if I didn't reduce myself to nothing and start again, I would never be able to see. The Austrian philosopher, Ludwig Wittgenstein, famous for being practically impenetrable, once described, with bracing clarity, the form of a philosophical problem as, "I cannot find my way," which describes, I think, life most of the time for most of us, including me. I found that New York was generous in that way. It allowed me to see again, to try and find my way, because it was continually doing the same thing. New York seemed to welcome questions.

After my attempt at romance atop Rockefeller Center, I turned to smaller pleasures. I would watch thunderstorms from the roof of my apartment building, although in New York the sky is often viewed as little more than undeveloped real estate and a thunderstorm—or any weather that slows traffic—is often perceived as a sort of personal affront. I have always loved thunderstorms. Once, when I was about five years old, I ran from my house into the backyard during a storm, face and hands up to catch the rain. A bolt of lightning struck, so close it blinded me for a few seconds, and the concussion enveloped me like a feather comforter, pinning my arms to my sides and tipping me over into the puddled grass. The air smelled like a hot iron. I was startled, but unafraid, since, at five, I knew I was immortal. Since then, my sense of immortality has eroded a bit, but when I watch from the rooftop, I am still delighted and I still hold my face and hands up to catch the rain.

Another of my smaller pleasures was reading the dialogues of Plato. At one time I had been required to read Plato, but that was at a military college where quantity was critical. We breezed through the whole of Western philosophy—and civilization in general—in a one semester course. In the end, the answers to perplexing questions such as, How should I live? What should I do? What am I meant to be? was, as with so much else, "Yes, sir." On the second reading, that answer seemed inadequate, and I was able to see some of the beauty of Socrates, who met my questions with more questions, and then even more questions. Slowly, he broke down my opinions, especially those I held about myself.

My favorite dialogue is the Meno, which begins with Socrates being asked, Can virtue be taught? The questioner, Menon, a wealthy young nobleman—today, he might be a Kennedy—begins with the assumption that he knows what virtue is, but Socrates presses him to define what he means by virtue, before he asks if such a thing can be taught. The dialogue ends with Menon confused. Through such questioning, Socrates felt one would eventually find answers, although they might not be the answers one expects and there might be only a moment of understanding—an instant during which one is, to use his metaphor, dazzled by the light. He warned that if one's vision—of the eye or of the mind— was perplexed and weak and one could not see, we should not be too quick to laugh. The person may be going into or coming out of the light, lacking or having achieved understanding. Both states are confusing.

One steamy summer day, in search of smaller pleasures, I stepped into the Northwest Coast Indian room at the American Museum of Natural History. It was cool and dark and peaceful. On each wall were the massive, carved wooden totems, which are unique to the Native Americans of the Northwest. I walked down the middle of the hall, as if on a forest path, dwarfed by the wooden columns, while the carved faces in the wood glared down on me. I sat and began to draw. I found that—and this is true of drawing anything, really—I was no longer a passive spectator, standing and gaping at their size and fantastic appearance. Instead, I followed the sensual curves, felt the shapes blend together and, at times, practically explode out of the wood. It was there that I first saw a killer whale, in a pole, or rather a section of a pole, displayed against the west wall, an abstraction three feet in diameter. I had to read Hilary Stewart's *Indian Art of the Northwest Coast* to decipher the shape, but even before I had precisely identified it, I enjoyed the form, the old and weathered wood, the bulging eyes, the long snout, and the second animal crawling out of its mouth, which looked like a frog. As I sketched, I noticed that a man's face was engraved in the forehead, at the spot where the whale's blowhole might have been. Man and killer whale were frozen together. I studied several other carvings. I followed the contours, the flow of the lines from one figure into another, like a dense web. Where did one creature end and the other begin? It was many, but it was one and it smelled like rain and the air after the lightning has struck.

If the carving itself was a web, it had also once been part of a similar web, which had been shredded. The carving had then been placed in a museum for me to observe for a few seconds on a summer afternoon. The killer whale I admired had been scooped up during what has often been called the "Museum Age," which could just as easily be called the age of confiscation. At the entry to the Hall of the Northwest Coast Indians, on the left, a plaque identifies the anthropologists of the "Jesup North Pacific Expedition, 1897–1903," who collected the carvings and other materials, an expedition made possible, according to the plaque on the right, by the "munificent gifts" of Morris K. Jesup, although I would later find that many of the carvings were obtained long before and after the expedition.

Jesup was a financier who had become a multimillionaire through a railroad supply distribution business in the mid-1800s. He later served on the board of the American Museum, which had been established in 1869 for the purpose of "recreation and education . . . upon a scale commensurate with the wealth and importance of our great city," a bit of chest thumping by wealthy people. The thinking was, Washington, D.C., had the Smithsonian Institution and Boston had the Museum of Comparative Zoology, so New York, the richest city in the country, should have the American Museum, which would take the same central importance in American life that the British Museum had in England. During that same period, New York diarist George Templeton Strong, businessman, lawyer, and philanthropist, wrote in his diary—December 19, 1868—that New York government was "diseased" and "corrupt" and a community "worse governed by lower and baser blackguard scum than any city in Western Christendom, or in the world, so far as I know." Jesup, who became museum president in 1881, knew how to deal with blackguard scum. He provided jobs to the constituents of politicians connected with Tammany leader G. W. Plunkitt. Consequently, the city increased its contributions to the museum from less than $1,500 per year before 1880 to $120,000 per year at the turn of the century. In 1877 the first building was completed at a site called Manhattan Square, just off the west side of Central Park. Five more wings and a completed south facade would eventually be added.

NEW YORK

It was a triumphant space and it had to be filled. Jesup's tenure coincided with what historian Douglas Cole has described as the "scramble" for Northwest Coast Indian artifacts, which was part of a larger, restless desire on the part of European countries and the United States to bring back and put on display every plant, animal, and artifact that wasn't securely nailed down. The British Museum was stuffed with artifacts from its empire. German museums were full of material from East Africa. Belgium had its own Congo Museum. There was so much collecting going on that Spencer F. Baird, secretary of the Smithsonian, became concerned about foreigners carrying away Native American materials by the shipload and wrote, in 1882, without a trace of irony, "I wish there was some law that prohibited foreigners from coming in and carrying off all our treasures."

Native peoples were also caught in the roundup. In 1885, nine Native Americans of the Bella Coola group from the Northwest Coast were signed up by a German collector to tour for one year. They were paid twenty dollars a month, plus food, lodging, clothing, and medical expenses. In Berlin, they gave a special performance before the Society of Anthropology, Ethnology, and Prehistory, and thus Franz Boas, later of the American Museum in New York, saw his first display of Native Americans and the foundation of his future career. He studied their language, legends, and music for several weeks. Forty-seven different physical measurements were made of each Indian, including "cephalic indices"— head size—and the height of their symphyses pubes—the penis.

Enormous exhibitions of native peoples followed in London, Paris, and in 1893, Chicago, with the World's Columbian Exposition and its Midway, a mile-long strip of land containing model villages of Dahomaians, Javanese, Egyptians, Samoans, and Native Americans, with nearby restaurants and a ferris wheel. As part of the Midway, Boas, assistant to the fair's chief of archaeology and ethnology, arranged for fifteen adult Northwest Coast Indians and two children—Kwakiutl, one of seven Northwest Coast groups—to come to Chicago, where they lived for the duration of the fair in a large native house painted with a thunderbird and moon images, which had been transported from British Columbia. During the fair, "Midway Types: The Chicago Times Portfolio of Midway Types" was published, which described Native Americans as "monotonously hungry to kill somebody, a white man if possible, another Indian if the white man is happily absent."

During this worldwide race to acquire and display everything, the American Museum, with the help of Cornelius Vanderbilt, John D. Rockefeller, and J. Pierpont Morgan, obtained thousands of pieces of Northwest Coast native art and artifacts, including the sixty-four-and-one-half-foot canoe, with the killer whale and raven painted on the prow, that now sits in the lobby in front of the Northwest Coast room. Thus, by the time of the 1897 Jesup expedition, the American Museum already had quite a collection of Northwest Coast art, as did the Field Museum in Chicago, the Smithsonian Institution, and British Columbia's Provincial Museum, which was established in 1886 because foreign collectors, mostly American and German, were "depleting" the province of native artifacts. The Field Museum in Chicago announced a similar expedition, so even at that relatively late date two American museums were roaming the coast, competing with one another for art, artifacts, and bones, an institutional rivalry that brought a carved killer whale totem to New York, a piece of the exotic put on display, everything being rounded up on the principle that the world belonged to us. The objective was, of course, scientific, described by Boas, who by then was assistant curator of anthropology at the museum, as an investigation of "the history of man in a well defined area." They would collect artifacts, myths, stories, linguistic information, and measure the physical characteristics of the natives, by using accepted scientific methodology that included robbing graves. As Boas wrote home to his wife during a trip to the coast, stealing bones from a grave was "repugnant work" and even prompted horrid dreams, but "someone has to do it."

Having been torn from its web of relationships, the killer whale is now anonymous. The pole I looked at and drew, and looked at and drew again, until I could almost feel the hands of the carver on the wood, is not identified at all. I wrote to the curator at the museum and asked if he could identify it. An assistant responded that it was Haida, collected during the Jesup expedition in 1901 in the Queen Charlotte Islands. While Boas had insisted on thorough firsthand field research, on getting native explanations, on attention to the play of the imagination and consideration of the artistic process, the closest one can come to these sculptures having a name is "collected by." The pole I drew was collected by C. F. Newcombe, which I determined from looking at published photographs taken at the time. A person carved it, though, and I often wondered if the carver of the killer whale looked over at the work of another carver and said aloud what he thought of his work, just as the Romanian-born sculptor Brancusi, known for the minimal, smooth form of "Bird in Flight," is said to have referred to Auguste Rodin, sculptor of the bulging muscles of "The Thinker," as a "maker of beefsteak." No history accompanies the carving, though, no school of art, no blue period of abstract expressionism for the carver, because they are objects in a museum and the museum is our tradition, not those whose carvings were collected.

Lacking identification and their own periods of style, however, did not keep them from eventually finding a place in the European schools of art. In 1946, as Edmund Carpenter explains in the book, *Indian Art of the Northwest Coast,* Max Ernst, a refugee from Europe and a surrealist painter, and Barnett Newman, a young American abstract expressionist, helped mount an exhibition of Northwest Coast Indian paintings, with pieces brought together from private collections and the American Museum. Newman, in the introductory essay for the exhibition, wrote that the works were an answer to all those who assumed modern abstract art was "the exercise of a snobbish elite." Among the Northwest Coast Indians "abstract art was the normal, well-understood, dominant tradition." So, in midcentury, the collected objects were recognized as art in order to help justify abstract expressionism, a sort of, as one historian described it, "aesthetic equivalent of decolonization." Carpenter writes, "By moving the Museum pieces across town, the Surrealists declassified them as scientific specimens and reclassified them as art." Perhaps the artists felt they were doing them a favor, in the name of art, just as the American Museum of Natural History felt it was doing a favor, in the name of science, by collecting the objects—the last embers of a dying race—in the first place.

Just three years before that exhibition, French anthropologist Claude Lévi-Strauss, well known for his books *The Savage Mind*, the central theme of which is the mental equality of all humans, and *Tristes Tropiques*, wrote an article while he was living in New York City, which is quoted by nearly everyone who makes even passing mention of the Northwest Coast art in the American Museum. I won't be the exception. The article begins, "There is in New York a magic place where all the dreams of childhood hold a rendez-vous, where century old tree trunks sing or speak." He's right about that. He goes on to describe the exceptional feature of the art as "the almost monstrous faculty to perceive as similar what all other men have conceived as different." I believe that's what I was seeing as I drew the killer whale, a web of relationships that said we were not so different, not so far apart. I thought, then, that I would go looking for whales, to see for myself.

I began to see killer whales everywhere. On television, I saw two advertisements, each with film of a killer whale leaping out of the water. In one, for Dupont, the killer whale was followed by a shot of sea lions applauding, and the voice-over described Dupont's commitment to the environment. The other was about Princess Cruises. A killer whale leaped from the water, framed by blue-white glaciers and jagged mountains, and the voice-over said, "It's more than a cruise. It's the love boat."

A few weeks later, I had a dream. I was on a school bus in Guatemala with my girlfriend, Karen. Suddenly, the bus drove over a cliff. If you've ever ridden a bus in Central America, you know that the dream, up to then, was fairly realistic. I was trying to hold on to the bus, to Karen, and to a bag of Ann Taylor clothes, which is where Karen shops all the time—good sales and a good return policy, she tells me—and the bus was transformed into a killer whale's dorsal fin, which is the large, sweeping, triangular shaped fin that projects from its back. I held on tightly as the whale swam through the air toward the ground below, then began turning loops, as if it was trying to swallow its own tail. All the while, I was hanging on. It spun around and around, faster and faster, and it finally threw me off. I floated away, up in the air, still holding Karen's hand and the bag of clothes, as the killer whale continued spinning down and down and down. It disappeared. I woke up.

It was a good time to leave. I wrote a few letters, packed a rucksack, and looked at some maps. At a party before I left, over sourdough bread and cold beer, I told some friends I was going to look for killer whales and I wasn't quite sure how far I would go, although I explained that I'd seen a totem in the museum that had come from the Queen Charlotte Islands, and I thought I might eventually end up there. At least I was headed in that direction. They peered at me as if I'd said I was going to look off the edge of the world.

Then, late one night on Channel 11, I watched a movie called *Orca*, which is from the Latin term for killer whale, "Orcinus Orca"; Orcinus meaning "of or belonging to the realms of the dead" and Orca "a kind of whale." In the movie, Charlotte Rampling, who played Rachel, the sympathetic marine biologist, said Orca meant "the bringer of death," which was a slightly different shading. *Orca*, the movie, was released not long after the movie *Jaws* and was trying to capitalize on the latter's popularity by substituting a killer whale for the great white shark. The television station had paired the two films. *Jaws* was playing later in the week and was advertised during commercial breaks ("Jaws—Uncut—So You

Won't Miss a Bite"). Richard Harris played Captain Nolan, an Irish fisherman with a crusty temperament, who harpooned a pregnant Orca while trying to capture it. The baby Orca was killed. Nolan released the wounded female, which, followed by its mate, swam off and died. The male decided to seek revenge. He rammed the docks, which caused a fire and practically destroyed the entire village, then bit off Bo Derek's leg and, in general, raised quite a ruckus. Eventually, Captain Nolan understood the destruction to mean he was supposed to go out and face the whale. That he did, for some odd reason accompanied by Rachel. He followed the whale on an eternal trip through a freezing northern sea, until they were trapped amidst sparkling white glaciers. The boat was completely encrusted with ice. After a certain amount of banging and jockeying for position, the whale, with a quick flip of his tail, tossed Nolan against the ice and killed him, but spared Rachel. The captain's last, gurgling words were, "What in hell are you?"

There are more than eight thousand animals on display at the Vancouver Public Aquarium, but only one warrants a headline when it dies. A few months before I arrived, Hyak, one of three killer whales at the aquarium, had died of a severe pneumothorax—a perforated lung. He had been captured in British Columbia waters at the age of two and his death ended what the *Vancouver Sun* described as a "23-year show business career." The headline read: "Hyak Buried at Sea, Might Leave Heir." Bjossa, the sixteen-year-old female on display, was showing pregnancy levels of progesterone. If she delivered in six months or so, Hyak would be the deceased father. The two whales had previously had a calf, in 1988, the first born in captivity in Canada, but it died after just twenty-two days. Hyak's death prompted dozens and dozens of letters from children, most of them with enclosed drawings, all of which were very similar to this: "Hyak was the first killer whale I had ever seen. We went on a camping trip and went to see the famous Hyak. I loved him so much. I am sorry! Love, Brodie. Age 10."

Within six hours of his death, Hyak was removed from the pool by crane and transported by flatbed truck to a barge in the shipyards of North Vancouver. There, he was laid out under floodlights and surrounded with tubs of formalin, a preservative. The postmortem on an animal of Hyak's size—twenty-three feet six inches long and weighing 13,358 pounds—is less like surgery and more like tearing down a building. A team of six veterinarians and pathologists, along with other members of the aquarium staff, assembled to perform an autopsy. The team opened the carcass, peeled back the abdominal wall, and examined the respiratory system and intestinal tract. Then, one by one, they began removing organs and weighing them. When an animal in a zoo is sick, everyone in the network of zoo professionals knows about it. Because so few such animals are available for study, especially large mammals, there is a long wish list. In Hyak's case, a geneticist wanted the retina; others were interested in trace minerals and pesticide accumulation in the blubber (nine and one-half inches thick); a cardiologist wanted the heart (thirty-four and one-half pounds); the Naval Ocean Systems Center in San Diego wanted the brain (fourteen and one-half pounds). Dr. Ron Lewis, the assistant chief veterinarian of the British Columbia Ministry of Agriculture and Fisheries Animal Health Center, who directed the autopsy, told me, "We had pieces going everywhere. It was like a scrapyard."

After eight hours of cutting, pulling, and preserving, only bits and pieces of Hyak were actually saved—the skull and spine being the largest bits. Samples were taken of virtually everything. Pesticide contamination was found, but at very low levels compared to samples taken from whales found stranded on British Columbia beaches. No parasites were found, which are a problem in wild whales, but they did find what in humans would be considered pre-Alzheimer's lesions. The lungs were significantly damaged, probably from a case of pneumonia in 1983, when an abscess had formed and been walled off with connective and scar tissue. For some reason, the abscess had become active and the infection had spread. The official cause of death was twofold: A perforation of the lung, less than one-half inch in diameter, had allowed air to escape into the chest cavity and collapse the lung; and a chronic inflammatory condition had developed because of scar tissue. Hyak was sick for three weeks before he died. He must have felt as if someone was sitting on his chest.

The remainder of the carcass was wrapped in canvas and weighted with five tons of chain. The barge was towed to a twelve-hundred-foot-deep area of the Strait of Georgia, the major waterway north of Vancouver, and a small bulldozer was used to push the carcass off the barge into the strait, while two of Hyak's trainers offered a farewell toast with Grand Marnier. Unfortunately, the chain went first. The canvas had ripped and a few days later small pieces of blubber and skin, carried by a strong southeasterly wind, floated ashore near Sechelt, not too far from where Hyak had been captured.

For the aquarium, that was an embarrassing end to a long

GONE WHALING

career. A Vancouver animal rights group called Lifeforce had already publicly questioned the speed with which the autopsy had been performed and the carcass disposed of, and when pieces started floating ashore it looked, well, at the very least, undignified. After fielding several complaints, the aquarium's general curator responded with a letter to the editor of the *Coast News* in Sechelt, in which he said he didn't know "what contribution to our knowledge of whales is served by so focusing on some pieces of blubber on the beach." He did not mention the intense interest in similar pieces shipped throughout the United States and Canada.

A few days later—perhaps even as Brodie was composing his letter—the *Vancouver Sun* headlined on page one, "Visitors Watch Whales Kill Trainer." Keltie Lee Byrne, twenty years old and a trainer at Sealand of the Pacific in Victoria on Vancouver Island, had slipped into the pool. Three captive killer whales had pushed her around and dragged her under water, apparently playing with her as if with a new toy, until she drowned. Ms. Byrne was a third-semester student at the University of Victoria studying marine biology and had been a member of the Canadian swim team at the Pan American games in Indianapolis in 1987. The week before, she had won the two-hundred-meter breast stroke at the Canada West University Championships in Edmonton.

The incident prompted an inquest and a flurry of public attention. In an editorial, the *Sun* said the tragedy underscored how "dubious" the "Aquarium entertainment industry" was and that "even at the better facilities like the Vancouver Aquarium, captivity takes its toll. Hyak, the whale that died there last weekend, aged twenty-five, might have lived to be fifty in the wild," although Dr. Lewis told me that if Hyak had remained in the wild, he would have died a number of years ago. He wouldn't have been able to fend for himself. Animals that get sick in the wild often don't survive long enough to get much sicker. In any case, the controversy over the death prompted the municipal council to announce that it would not automatically renew Sealand's lease on its valuable waterfront property. Shortly thereafter, Sealand announced it was closing down its killer whale program. The vice president of Sealand said it was due to the "gut wrenching death," but there was a certain amount of understandable suspicion that it was a simple cost-benefit analysis; the property was worth more than the whales. They eventually shipped the whales to Sea Worlds in Florida and Texas.

The Vancouver Aquarium is located in Stanley Park, Canada's largest city park, named for Lord Stanley of Preston, Canada's governor general in 1888, when the park opened. More than half the park's nearly thousand acres is forested peninsula, dominated by Douglas fir, hemlock, and western red cedar, which juts into Burrard Inlet, part of the busiest port in Canada. The aquarium, a nonprofit institution with a $6–7 million annual operating budget, is tucked away in the trees about a ten-minute walk inside the park and draws about 1 million visitors each year. When it opened in June 1956, it was a single building with a staff of four. Today, it's a large complex with, among other things, a Tropical Gallery, an Amazon Gallery, a British Columbia Hall of Fishes, pools for seals and sea otters, a beluga whale habitat, and a killer whale habitat. That exponential growth, necessary, according to an aquarium brochure, "to meet the Community's needs," coincided with the presence of Dr. Murray Newman, a Ph.D. in zoology from the University of British Columbia, who joined the aquarium at its inception and has been director since 1966.

While Dr. Newman was the aquarium curator, in 1964, a book was published entitled *Management of Wild Mammals in Captivity*, written by Lee S. Crandall, general curator emeritus of the New York Zoological Park. Based on fifty years of experience in keeping wild mammals in the New York Zoological Park, it contained information on elephants, spiny anteaters, bandicoots, wombats, elephant seals, hippopotamuses, and kudus. For example, the hippo, wrote Crandall, has had "a long career in captivity"—the first living specimens to reach Europe were kept in the menagerie of Roman Emperor Augustus between 29 B.C. and A.D. 14—and is "notable for its ready acceptance of captivity and for the excellent longevity records it has established." The elephant, he wrote, is "amenable in general to the restrictions of captivity." There was nothing in the book, however, on killer whales and perhaps that's why, that same year, Dr. Newman decided not to capture a killer whale, but to commission a life-sized model for the new British Columbia Hall. The sculptor he hired, Samuel Burich, needed a specimen. With an assistant, he rigged a harpoon gun on rocky Saturna Island, in the Gulf Islands of British Columbia. A few months later they harpooned a fifteen-foot, one-ton whale.

The whale survived the ambush. When Newman was told that the whale was still alive, he decided to try and save it. They towed it for twenty-six hours to Vancouver Harbor and put it in a makeshift pen at the Burrard Drydocks. A week later, it was moved to another pen at Jericho Beach near the University of British Columbia. That whale—named Moby Doll—was the first killer whale in captivity. For fifty-five days, night and day, she circled her pen counterclockwise and refused to eat. She didn't fight and she didn't try to escape. Once she began to accept food, she survived for thirty more days, then died, apparently from a skin disease caused by the low salinity of the harbor water. After her death, biologists determined that Moby Doll was a he. For his part in the capture, Dr. Newman was named Man of the Year by the City of Vancouver Visitors Bureau.

Because Moby Doll had behaved so gently, resigned or perhaps just in shock, his capture touched off a thirst for more. Between 1965 and 1973, forty-eight killer whales were captured in British Columbia waters and sold to aquariums and oceanaria. Twelve others died during capture. In 1967, the aquarium bought for eight thousand dollars one of seventeen whales that had been trapped in Yukon Harbor. They named her Skana and put her on display. One year later they had their second, Hyak. It was not a coincidence that the growth of the aquarium followed not only the arrival of Dr. Newman but the arrival of killer whales. Soon after that, the killer whale was made the aquarium logo.

I made my way through the fresh, dripping green of Stanley Park on my way to visit Dr. Newman, who had advised me over the telephone that he would appreciate a check for $20 million, which would help him with plans for the next expansion, a "Pacific Canada" exhibit. The rain had stopped, but it was only a pause. It was June and a forty-year-old record for the least hours of sunshine in a month was about to fall. For the first time in forty-two years, the Vancouver Rose Society was forced to cancel its annual show because of a "poor showing" by the flowers, which need four to six hours of sunshine each day to stay healthy. On my way to the aquarium, I spoke to several people at a Starbucks coffee bar on Pender Street. They were convinced that summer would never arrive. A few minutes later, those same people immediately identified me as an American when I looked both ways and walked against a "Don't Walk" light, leaving them behind on the curb. I had violated the Vancouver sense of order. I could feel them staring at me. Halfway across the street, stricken with criminal guilt, I looked back and apologized.

Once in the park, I stopped to read the *Toronto Globe and Mail* beneath a statue of Robert Burns (1759–96). His arms were folded, he was wearing breeches and a long topcoat, and he looked sternly over the boats docked by the Vancouver Rowing Club. He was made less stern by the seagull sitting on his head. I was unsettled by the following headline: NEW YORK BUDGET CUTS DEATH SENTENCE FOR POOR. The story said that a city program credited with reducing infant deaths in New York's ghettos was slated for extinction under proposed city budget cuts. The Infant Mortality Initiative cost $10 million annually and infant death rates had leveled off since the city launched the program six years ago. But, the article continued, the cries of protest for ghetto babies had been drowned out by screams to save zoo animals. Since budget cuts had been unveiled last month, politicians had been deluged with petitions from affluent Manhattan residents to reverse the closing of the Central Park Zoo, which cost the city $2 million annually. The article quoted Elizabeth Graham, New York's assistant health commissioner for maternity services, as saying, "I suspect the zoo stands a better chance of surviving than our infant mortality program."

As I stood there in the park, I couldn't help but think, What in the world am I doing here?

Dr. Newman probably thought the same thing. He may have thought it was about the worst idea he'd heard lately to entertain more questions from the curious. Referring to the poor publicity following the deaths of Hyak and Keltie Lee Byrne, the general curator told me they had been "badly battered." The receptionist raised her eyebrows in my direction and said that there had been an unofficial "no press" policy for several months, so neither she nor I could settle on a reason why he agreed to speak with me, unless he actually expected me to have that check. I knew, however, that he was not without a sense of humor. On the way to his office, I stopped at the H. R. MacMillan Tropical Gallery, named after a twentieth-century lumber baron and the aquarium's first large benefactor. Inside, there was a slightly less than life-size black and white blowup of Dr. Newman, wearing a Tilley outdoor hat, a plaid shirt, and field pants, standing by the entrance. He had a camera at his waist, a South Pacific Handbook—in color—at his feet, a color globe in his left hand, and he was pointing with his right to the South Pacific. Above his head, a blue-green idea bubble had him saying, "Join us. We're exploring coral reefs. The Vancouver Aquarium has been studying these incredible places for over 20 years."

Dr. Newman greeted me in the reception area and walked me to his office, where we sat on two brown couches near several large, floor-to-ceiling windows. A spotting scope looked out on a long stretch of the thick lawn of the park, which sloped gracefully toward Burrard Inlet. The Coast Range of mountains were visible, although the peaks were socked in by rain clouds. A Hyundai tanker was docked in the harbor. Dr. Newman had a round face with clear blue eyes. He wore charcoal gray slacks, a gray herringbone jacket, white shirt, and a blue and brown striped tie. He looked much better than the cardboard cutout, but he was nearly as reticent. He smiled reluctantly, like a man with bad teeth, even though his teeth were fine.

I mentioned that I'd read the recent criticism. He sighed. He rubbed his eyes. He appeared to be in pain. He said that most criticism of the aquarium was "sublimation," and it was easy to worry about killer whales, rather than about real problems. "People who come from Chicago or New York, they understand what real problems are. Ten thousand homeless people, that's a problem. Out here, they worry about whales in captivity." He offered the faintest of smiles and added, "Maybe it's something in the air."

"Why killer whales?" I asked.

"It's a very symbolic animal," he said. "At first, we wanted one because it's an animal that symbolizes these waters, as it did for the native people. I don't think it was necessarily in people's minds, but I wanted it as a symbol. So, our original concept was for a sculpture of a killer whale." He paused and rubbed his eyes, as if he'd told that story too many times before. "We think of the aquarium as essentially a living museum," he said. "Canada is in many ways wilder than Africa in terms of its wildlife population. I feel that zoos and aquariums are centers to tell people about these animals and to create a sympathetic regard for them. In the case of the killer whale, we have exhibited them since the first one was captured alive, and since then there has been a transformation in public attitude about the animal, from undesirable predator to magnificent wildlife species. We have played a part in that transformation. But, public attitude is fickle. Long-term conservation requires consistent, stable presentation in institutions like public aquariums." He paused. I waited.

"In Canada, history shows that animals have been utilized," he added. "An aquarium helps people maintain a level head."

He was interrupted by a telephone call. He seemed genuinely relieved to be on the telephone with a friend, talking about the superb diving in Papua New Guinea. After a few minutes, he concluded his conversation, but remained seated behind his desk.

"There's an immense sensitivity about animals, but there is also an immense sentimentality," he said. "The killer whale is of great interest to aquarium professionals because they adapt. They are stable in captivity. They are not frightened of people and they are completely at ease in their space. If you put them in a pool, they don't panic."

He got up, came over to the couch, and flipped open a scrapbook of newspaper clippings. It was a family scrapbook, but a family with nothing but bad news. Almost all of them were critical of Hyak's death or attempted expansion by the aquarium. Here's the photo of when Jimmy broke his arm and leg. Oh, here's Mary when she was sick with intestinal parasites.

"I think what we do is give the killer whale a boost in the priority of mankind," said Dr. Newman. "But killer whales in aquariums are analogous to elephants in zoos. Zoos are trying hard to have breeding groups of elephants, because they're focal points of interest, but there are great technical problems—it's dangerous, costly, and male elephants require special care. It takes a lot of motivation on the part of the institution to maintain them."

I suggested that perhaps it had all backfired on him; that if he hadn't played a role in transforming public attitudes, there wouldn't be any complaints now.

"Well, we're accustomed to these things being passionate," he said. "The large majority of people want to see a killer whale. A few articulate people question it and influence the thinking. Obviously, it's a community responsibility to see that they're properly accommodated." He paused for a very long time and rubbed his eyes again, as if he was trying to wait me out. I was patient. "There's no reason why we couldn't have a breeding colony, if we had adequate facilities," he finally said. "There's no magic number, but they interact, so it's good to have several. The ideal number might be four or five. In nature, you have mainly maternal pods, an adult female surrounded by her young. So, if you want to create the most natural environment, you would have a maternal pod. That's quite possible, if you have a good pair."

He tapped his finger on the scrapbook. "This is not the important stuff," he said. He explained his new interest in "biogeography." All the exhibits will eventually have a geographical focus and be coordinated with aquarium-led research and conservation. Also, he wants the aquarium to extend its reach into marine mammal rescue and research. In the past, for example, the first of its major ecological programs was concerned with the Amazon Basin, where a research associate was established in Manaus, Brazil. In 1983, the Graham Amazon Gallery was opened. A second regional project began in 1968, with the first research expeditions to Baffin Island to study narwhals. In 1990, they opened an Arctic Canada exhibit, with a beluga whale habitat and an interpretive gallery focused on Lancaster Sound in the Arctic. Now, in a further attempt at transforming a "zoological garden into a biological park," the Pacific Canada exhibit will feature an underwater habitat that will be modeled after the rocky shore of the Queen Charlotte Islands.

"How much do you need for Pacific Canada?" I asked.

"Six point six million," he said.

"How will you raise it?"

"Well, I don't know." He paused for a long time, as if he were lost in thought, then added, "I spend a lot of time thinking about it, though."

"**W**hy is the dorsal fin bent over like that?" is the most commonly asked question by visitors to the aquarium, and it refers to the fact that the male's dorsal fin is bent over to the left. The answer: Because the dorsal fin is connective tissue with a mass of blood vessels, but no bones or cartilage, and is used for heat regulation and stabilization underwater. In captivity, whales spend more time on the surface, with the fin out of the water. Unsupported by water, it eventually begins to droop.

The second most commonly asked question is, "Do you think that they're happy?" which is a far more difficult question. I heard it answered this way: "Well, we have no way of knowing if they're happy, but they are healthy and well adapted." When I heard that answer given, I wondered what was meant by the term well adapted, which Dr. Newman had also mentioned. The answer was not clear to me until I was shown a videotape of some narwhals, arctic whales that have a long tusk on their snout, who, after capture, had beat themselves to death on the side of the pool.

The "killer whale habitat" was completed in 1986 and received a Significant Achievement Award from the American Association of Zoological Parks and Aquariums (AAZPA) the following year. It holds 1 million gallons of water and ranges in depth from fifteen to twenty-two feet, in three connected pools. Twenty-five-foot sandstone formations, designed to look like the craggy shore of the Gulf Islands, rise from one side. On the other side rises a concrete bowl of seats for about five hundred people. The juxtaposition is an apt representation of a debate that was raging within the aquarium. On the sandstone rock side—the illusion of the natural—were those who felt the scheduled killer whale shows should be eliminated, so that people would come to watch the whales the same way they would anything else in the aquarium, although the trainers would still come out several times a day to "work" the whales. The message would be, these animals are important not because they can perform but "just because they are there." A spokesperson for the aquarium, Stefani Paine, described the new approach as the "kind of thing you could experience in the wild."

On the other side, sitting in the concrete bowl, were those who argued that the killer whale was not just any other animal, it was the aquarium's biggest attraction. Without a scheduled show, people would stop coming. They came with the expectation that there would be something to clap for and that was a difficult expectation to overcome. The debate was eventually won by the illusion of the natural side. This changed the routine for the trainers only slightly. They would still have to work and feed the whales a certain number of times each day, but visitors would no longer arrive at specific times to watch. The tricks themselves, though, had already been eliminated. In Vancouver, nobody rides around on the whales or puts funny hats on them to celebrate Canada Day. Instead, the killer whales have been trained to demonstrate "natural behaviors." The marine mammal staff is expected to provide the whales with mental and physical challenges, to replace those they might have had to face in the wild.

A few days after my conversation with Dr. Newman, and shortly before the new plan was to take effect, I was trailing after one of the trainers, Doug Pemberton, trying to get some idea of what he did, which apparently included a bit of everything, unlike other oceanariums, where trainers might simply work with the animals. "Some people think, well, you come out five times a day and feed the whales, that's the job for me," Doug said. "Then they quit after two days, right after you tell them to hose down the pool. They think, great, feed the animals. They never think, great, clean up after the animals."

Doug had no such illusions. He started working at the aquarium twenty years ago "ankle deep in seal shit." He came to an interview in a coat and tie hoping for a summer volunteer job and at the conclusion of the interview was asked to start cleaning out the pool, which he did. Within a few years, he was training killer whales and, for many years, worked exclusively with the famous Hyak. When he first touched a killer whale he said, "Yeah, that's it," just what he expected, rubbery, like a wet suit used for scuba diving. What was unexpected was their intelligence. After a few minutes with them, he discovered that training wasn't simply a case of giving them a cue and watching them perform. They didn't just run off and do what he told them to do, but often went off and did something else or seemed to be acting completely on their own. It was less like training and more like learning to understand the moods of a sometimes cranky uncle.

Doug is about five feet six inches tall with a slight paunch. He's balding, has a closely trimmed beard and mustache, and blushes easily. It was midmorning. We were beneath the main pool in a tiny room that smelled like a flooded basement. He was wearing navy blue Helly Hansen rubber boots and a royal blue Vancouver Aquarium sweatshirt. I had already watched him drain the research pool, so he could put a seventy pound seal—which had a small bald patch from an electrocardiogram—in a cage and manhaul him out. Then, he tossed fifteen pounds of geoduck, a northwest clam, to the three sea otters who had been rescued after swallowing a few of the 11 million gallons dumped by the Exxon Valdez, when it ran aground in Prince William Sound, Alaska. When the otters arrived, eighteen months before, they immediately began throwing tiny rocks against the fifteen-thousand-dollar windows of the brand new Earl Finning sea otter pool. Doug went snorkeling in the pool to get the rocks out. One of the otters playfully reached for his snorkel. Doug playfully pushed him away. The otter, less playfully, carved up Doug's shoulder, arm, and finger.

The Marine Mammal Food Room was next door to the trainers' room. With four industrial-size stainless steel sinks, it is filled with the sound of continually running water. Doug filled several stainless steel buckets with just-thawed herring. Each day, Finna, the male killer whale, is fed 175 pounds of herring; Bjossa, the female, 120 pounds; and Whitewings, a Pacific white-sided dolphin, 20 pounds. The aquarium buys 110 tons of herring each year from local fishermen, at $650 a ton. It's the leanest and the easiest to freeze. I was told that the quality of the herring is much higher than anything that I might buy in a store. "It's crap what they sell to people in the store," a trainer said. "I'd throw it in the garbage and I certainly wouldn't feed it to the whales." The whales have become accustomed to good herring. After her steady diet of it, Bjossa had to be trained to eat other kinds of fish. If a trainer tried to sneak a twenty-pound salmon into a bucket of herring, Bjossa would swallow it, but a few seconds later— "yeeach"—out would pop the salmon. Once they learn to eat something, killer whales seem to stick with it. One trainer theorized that a killer whale has never attacked and eaten a human because, when they see a human in the water, they don't recognize it as food. The drawback to using frozen fish, however, is that all the water soluble vitamins—about two thirds of all vitamins—are gone. "Sea World Marine Vitamins" must be added. Finna, for example, gets the equivalent of forty-five human multivitamins each day, heavy on B_1. Until a year and a half ago, human multivitamins were used, geriatric ones, with added iron. If live fish were used, the vitamins wouldn't be needed, but live fish carry parasites and the whales would have them, too. Whales in the wild are full of worms.

Doug, two full buckets of herring in hand and a whistle around his neck, made his way up a short, damp concrete tunnel to the pool area. A memo from the general curator was tacked to the door of the ramp, dated February 27, 1991. It said, in part, in the interest of staff safety and Workman's Compensation Board Requirements, "From this day forward no one is allowed at the edge of the killer whale pool alone." That was as close as anyone came to mentioning the death of Keltie Lee Byrne at Sealand of the Pacific while I was at the Vancouver Aquarium.

At the top of the ramp, Doug passed through a wire mesh gate and a six-foot entry, which opened to the back pool area. Earlier, I had seen the entry from the concrete bowl. It looked like the entry to a cave. Doug was followed by two other trainers: Clint, tall, curly haired, and mustached, would be working with Bjossa; and Hans, thin, with long blondish hair, would be working with Whitewings. Clint made his way to the back, beneath the sandstone rockface, Hans to the front of the larger pool near the spectators.

GONE WHALING

Children had started to line the Plexiglas rail that surrounds the pool, pulling the hoods of their pink and blue raingear over Montreal Expos and Toronto Blue Jays baseball caps. A tiny blonde-haired girl sat by the rail and opened an orange and gray striped umbrella. Next to her sat an orange-robed, white-turbaned Sikh.

Finna and Bjossa, shiny black and breathing through the blow holes in the top of their heads with mild "kwoof" sounds (caused by release of pressure, like the sound of a bottle of seltzer being opened), were bobbing in the back pool as we left the cave. Both were sixteen years old and approximately eighteen feet long. Finna weighed more than three tons; Bjossa about two and a half. They were average size for killer whales and still growing. A male will have a mature length of about twenty-three feet (about the length of Hyak); a female less than twenty feet. The longest on record is thirty-one and one-half feet, small compared to the sperm whales or blue whales, which can be fifty to ninety feet long. Whitewings, thirty years old, but only six feet long, darted around and between them, flashing like a silver coin beneath the surface. Both whales were tilted to the side, watching with one eye, since, like birds, their eyes are located on the side of the head.

As he left the cave, Doug raised his right arm, cocked it at the elbow, and made a fist. Finna submerged, passed through the opening into the larger pool, circled the pool underwater, and breached—a leap into the air frequently seen in the wild—sending a huge splash washing over the Plexiglas barrier at the spot where "Splash Zone" was etched and where the children had gathered. In the meantime, Doug made his way to the center island. He left behind one bucket of herring and carried a target pole, a long white stick with a ball on the end, which made him look a bit like a circus impresario, while the delighted screams of the soaked children reverberated in the tunnel.

"Nothing is forced," is the cardinal rule among trainers, although captivity is forced enough, I suppose. Still, certain behaviors must be formed in order to properly care for them. The first step is to get them accustomed to people. Second, get them accustomed to eating dead fish. After that, the trainer will try to pair behavior with something other than food. At the Vancouver Aquarium, it's a whistle. With killer whales, however, before a trainer will touch them directly, the target pole will be used. The training sequence might go something like this: First, get the whale to accept a fish. Once it accepts the fish, introduce the pole. When offering the fish, place the pole gently in front of the head, close enough so the whale can't avoid touching it. As soon as the whale touches it, blow the whistle and toss him a fish. Eventually, the pole and the whistle will be associated with feeding. The whistle, or any sound (other marinelands use an underwater beep), becomes what is known as a "bridge," because it will fill the time between the correct behavior and the eventual reward, like herring.

After that, the pole is used to form other behaviors. At first, the whale might follow the pole around in the water. Simple. The trainer might then put the pole on other parts of the body. The next step, for example, might be to get the whale to bring the pectoral fin—the fin on the side of the body—up to the pole. That movement, as with all movements, can then be paired with a hand signal. Eventually, the pole can be discarded and hand signals alone can be used and continually "faded," perhaps becoming imperceptible to the spectators in the bowl. In the past, I learned, big hand signals were used, the "circus type," with arms and hands flying around. (These are still used at some marinelands.) Today, instead, a trainer might use only a shake of the head or a movement of the feet. Whales pick up other clues as well, and they often reflect a trainer's mood. If the trainer is enthusiastic, they leap high in the breach. If the trainer isn't feeling well or isn't too happy about being at work that day, the whale might respond with indifference, barely exiting the water in the breach, a way of saying, I'm not too excited about this arrangement either. Whales in captivity are students of body language.

The easiest way—the shortcut—to train a captive whale is to put it in with an older captive whale. The younger will mimic the older, like a neophyte watching an older vaudevillian. Finna and Bjossa, who were taken captive in 1980, learned the basic commands in about four months, about half from mimicking the veteran Hyak, half from training. By that time, though, Hyak was a ten-year veteran and wasn't particularly interested in them when they arrived. The trainers speculate that it was because Hyak and Whitewings were very close. "They were just happy together, no other way to describe it," one trainer told me. The two new arrivals were kept in a separate pool for the initial training. After a few weeks, the mesh gate that separated the two pools was raised. Whitewings careened in, nosed around, bit the two whales, and basically showed them who was boss. Hyak swam in, took one look around, and left.

The gate was left open for two days before Finna and Bjossa swam into the larger pool. Despite the slow start, within two or three weeks, Bjossa was the boss, as seems to be the case in all killer whale groupings, which are centered on a dominant, maternal female. It gradually became clear to the trainers that Bjossa had to eat first. If they tried to feed one of the others first, they'd turn their heads. Sometimes, Bjossa would trick them. She'd take a mouthful of herring, but not swallow it. Then, Hyak and Finna would eat. Once they'd finished, she'd show them she hadn't yet eaten and the two males would get very agitated. Perhaps they knew they'd been rude, or perhaps because Bjossa threatened to wallop them if it happened again.

       After his breach, Finna surfaced near Doug at the center island. The pole was being used not for the look of the big top, but because Doug, accustomed to working with Hyak, was relatively new to Finna. Also, Finna was just being introduced to variable reinforcement—which means that the reward would not always be the same. It could be one fish, or a bucket of fish, or a good scratching—and it was being introduced in small steps. If things got touchy or if Finna was confused, the simple way to clear up the confusion would be to put the ball of the pole on the surface of the water.

Doug crouched and scratched Finna's head. Finna blew—"kwhoof"—and opened his mouth. Doug grabbed Finna's tongue, holding it in one hand, scratching it with the other, as if he were rubbing out the kinks in a mass of pink sausage. He released the tongue, stood up, tossed Finna a handful of herring, then stuck his arm out to the left, hand open. Finna turned away and circled the pool on the surface of the water, making a noise like a variable car alarm—"reeooo, reeeooo, reeeooo"—and returned. Doug fed him three herring—one by one—and Finna submerged and offered his tail, which Doug examined. Finna surfaced and was rewarded with another herring. Then, with both hands, Doug grabbed Finna's tongue and shook it back and forth, as if he were pulling on a stick in a dog's mouth. Suddenly, he released it and Finna flipped upside down in the water. Doug ran his target pole along the pearl white underside. Then, he stuck his arm out to the right and Finna circled the pool while slapping his tail on the water, a movement called a tail lob.

Up above, in the bowl, the interpreter, in a white and aqua shirt with the aquarium logo (a killer whale on a wavy sea) introduced himself and asked if anyone could tell the difference between Bjossa and Finna. After pointing out the bent dorsal fin, he added that killer whales needed a lot of variety. "They need change . . . and so do we. Everything changes here. The wild changes and so does the aquarium."

"How many heard this would be a show?" he suddenly asked.

From the display of hands, about one quarter of them.

"Well," he said. "It's really not a show. This is something for the whales. They don't do tricks here. They're asked to do a lot of natural behaviors and every time the trainers work with the whales they'll ask them to do different things."

The two other trainers were putting their charges through similar routines. The trainer can have the whale do as much as he wants or nothing at all, but everything they are trained to do has either been observed in the wild or is necessary for the trainer to monitor their health. Is the whale being sluggish? Jumping high? Engaged? Responding to cues? For example, Doug told me that Skana did "shows," back in the days when that's what it was considered, up until five days before she died. Nobody even noticed she was sick until two days before she died. On the other hand, he said, Hyak "was always a bit of a wimp," lethargic with labored breathing when he was sick. Hyak tipped off the trainers to his illness by playing with his food, which prompted an immediate blood check (routinely done once a month, although Bjossa's was being checked once a week because of her pregnancy). His white cell count was high and for several weeks they sorted through various possibilities, treating the infection and searching for clues, but there was nothing to suggest that he was any sicker than he had ever been. "It's like if you have people who don't have an appetite—you don't think they're going to die," said Doug.

Thus, each session with the whales is a medical examination. When Doug ran his hands over Finna's tail, he was checking for hot spots, which might be a sign of infection. When Finna exhaled and cleared his blowhole, Doug was smelling it. If it smelled badly, it could be a sign of trouble. At the moment, Finna was showing evidence of a low-grade infection and receiving 240 tetracycline capsules of 250 milligrams each, twice a day. Thirty days of tetracycline should knock out the infection, but the dose must be carefully monitored for side effects. Killer whales are voluntary breathers, which means that, unlike humans, breathing is not an automatic response. They must be conscious. So, for them, sleep is more of a steady dozing, in which they nap for a few minutes, then awake to take a breath. Antibiotics could cause a fatal loss of consciousness.

A fourth party joined the pool session. A great blue heron left the Douglas fir behind the sandstone facade, where seven of them nest, and in a long, slow circling glide, landed on the concrete just a few feet away from me, then stood with the dignity of a tall English butler. He took light, precise steps, bending head and neck forward, then side to side, peering past me down the tunnel as if looking for a relative arriving on the red-eye flight from Los Angeles. It was an elaborate ruse, though. The heron had his eye on the bucket of herring that Doug had left behind. Just then, Doug tossed the remainder of a bucket of herring into Finna's gaping mouth, then stuck his arm out to the right, fist up. I knew it was the signal for a breach. The children in the crowd had no idea, but after the last breach, in preparation for the moment, they had moved into, not away from, the "splash zone," tightening the hoods on their raingear as if preparing to be set adrift in the Pacific. Doug started to make his way back to the cave to retrieve the second bucket. Just before he arrived, the heron hopped to the bucket and neatly speared three herring, one by one, as if pocketing canapes from a banquet table. Then, Finna breached, the wave washed over the children in the first row, and the cave echoed with their screams. I could see them soaked, with a look that said, Mom won't be happy, but, boy, I am happy. The heron, disturbed by the commotion, heavily took flight and settled just as heavily on the center island, where he hopped up and down, contending with the waves from the breach.

Doug, perspiring, took a handful of herring from the second bucket and waited. Finna surfaced immediately in front of us, mission to douse the children accomplished. He looked first with his round right eye, then leaned back and opened his mouth, a cavity that was pink, soft, and big as a dumpster. I thought of my own puny mouth. I opened it and tried to put my fingers in, just to remind myself of how small it was. The base of Finna's tongue opened and closed against the roof of his mouth, which was dotted with what looked like black ink spots. Doug tossed in a handful of herring. The

VANCOUVER

tongue dropped from the back of Finna's throat, a hole opened, and the herring disappeared. Ten more hefty hand- fuls. They vanished, as if sucked down a drainpipe. Finna rolled over and looked again, this time with his right eye.

"He probably wonders who you are," Doug said. Finna ex- haled—"kwhoof"—and his breath smelled chlorinated. Doug kneeled down and slapped the side of the pool. Finna rolled over, then backed up and opened his mouth again, as water sloshed over our feet. Doug emptied the rest of the bucket, stood up, and backed away. The herring disappeared down the drain.

"I used to work exclusively with Hyak," Doug said, as we watched the herring disappear. "We really got to know each other. We'd play games with each other. Reading the animal is pretty important. He knew I wouldn't take any guff, and he knew what a good show was. He always tested you. Finna, here, he might make just a half circuit of the pool or do a small jump and he thinks his day is done." Finna was bobbing up and down just a few feet from us, backing away, giving a long look with his right eye. He was probably not nodding his agreement with Doug's assessment, but he looked like he was. Seagulls circled above, barely visible against the overcast sky. The heron, with a slow-motion leap, rose into the air and flew to the Douglas fir outside the bowl. In the background, the crowd was drifting away, and I could hear the interpreter encouraging them to go below to the underwater viewing area, "a vantage point you would never have in the wild."

"Hyak was a surprise right until the end," said Doug. "Some very complex behavior he'd learn right away. Like the flipper splash while traveling around the pool. I set aside a month to teach it to him. He was doing it in four days. Some other behaviors, I'd think, I sure wish I could get him to do that, and bam, he'd do it. Other times, with some very simple behavior, he'd take forever, as if it was just too simple for him." He paused for a moment. "I enjoyed working with him. He enjoyed working with me. At least, I think he did."

He said that wistfully, as if he wished it did not have to be forever tinged with doubt, then added, "I didn't think I'd be able to go to the autopsy, but I forced myself, and after the first cut, it wasn't Hyak anymore." He stopped, trying to think of something more to say about a creature with whom he had spent each working day for fifteen years.

"Hyak was just neat," he said. "It was like being married. I really miss him."

I descended a flight of steps and a ramp to the underwater viewing area, the H. R. MacMillan Gallery of Whales, followed, thunderously, by Mrs. Bowman's class from a nearby elementary school. The gallery had twenty-four acrylic windows, each three inches thick and about six feet high; the carpet, wall, and ceiling were Santa Fe green. The sunlight, filtered through water and windows, softened the room's edges. The children quickly filled most of the windows. I made myself comfortable on the ledge of an empty one. A young girl, maybe ten years old, was standing in the next window, surrounded by several friends. She had light brown hair, bangs, blue shorts, and a white T-shirt and was turned toward her friends as Finna drifted by underwater, big as a house. Finna jerked his head toward the window, mouth wide open, as the girl turned and found herself face-to-face with forty-eight conical-shaped teeth. She paused for a millisecond. Then, she screamed. The other children joined her in a screaming chorus, a flock of sparrows picking up a danger cry. I nearly went through the ceiling.

After recovering, the girl turned to one of her friends. "That was less than an inch away," she said.

"Whales can't break through," her friend reassured her. "It's too thick." She paused, then looked doubtful. "I think I'm right."

Finna drifted away, then returned, followed by Bjossa and Whitewings, joined in a ballet, twisting back on each other, touching heads. The dolphin slid along Bjossa's length and glided over her tail, while Bjossa turned resolutely in a circle. From above, the whales had looked sleek, shiny, and black. Underwater, they looked grayish, mottled, and scarred. Every scrape and nick, every bite and scratch, was visible. Underwater, there was no Gulf Island backdrop; instead the environment resembled a swimming pool. As the whales drifted away again, I tucked myself further into the window frame and imagined drifting with them, breathing through the hole in my head. I could see the children at the windows screaming and making contorted faces, and I pushed away through greenish water and shafts of pale sunlight. A few powerful thrusts of my tail took me to one end of the pool; a few thrusts more brought me back to the window. I felt cramped.

While I was drifting around the pool, a young boy, perhaps eight years old, joined the girls at the next window. He was squirming with enough nervous energy to launch a rocket. He pointed as one of the whales drifted by the window.

"Look," he said. The four girls looked.

"I saw his"—he dropped his voice to a whisper—"dick sticking out."

The nearest girl gave him a withering look, an older sister look. "You did not," she said scornfully.

"I did," he insisted.

"You did not," she repeated firmly. "That's the female. Look at the fin."

He didn't look at the fin, because the stake she had just driven through his heart had pinned him to the wall. He looked at her helplessly. He was trying to find some retort that would enable him to wriggle free. I found myself inside his head just as easily as I had been inside the whale's.

The four girls sensed failure. They quickly joined in a chorus, "It's a *fee*-male. It's a *fee*-male. It's a *fee*-male."

He was blinking fast, trapped in a maelstrom of taunts. His sister—it had to be his sister—stepped up to finish him off.

"And she's going to have a *baby*," she said.

"In a way, the animals come last," wrote Heini Hediger, former director of the zoo in Basel, Switzerland, and author of several books, including *Studying Animal Behavior* and *Wild Animals in Captivity*, which became known in the United States as the "zoo bible." He did not mean that animals should not be well cared for, nor live under the best possible conditions. Hediger believed zoos were extremely important as a last refuge for endangered species and as a "substitute biotope" for the city dweller. He also recognized, however, that the zoo did not exist for the benefit of the animal, regardless of how we might try to present it. A million-gallon killer whale habitat with sandstone backdrop is undoubtedly better for the whales; nevertheless, it is also a million-gallon stage set, for the benefit of the spectators, not the whales. In that sense, the message presented by the aquarium collides with its changing philosophy. If the philosophy is part of a "profound movement in having the whales appreciated for what they really are, and that is a magnificent animal," then what are they doing here? The marine mammal staff is described as being there for the whales' benefit, which the aquarium believes is a "discrepant event," something that challenges people's preconceptions and helps them make the shift from a human-centered to an animal-centered perspective. Yet, the killer whale must be trained to perform "natural behaviors" on cue. It is one of the ocean's great predators—scientists have examined the stomach contents of killer whales aboard whaling ships and found the remains of nearly every kind of whale, including the sperm whale and the blue whale, the largest creature on the planet—but they are not shown hunting and eating. The killer whale has been trained to eat frozen fish by the bucketful. Among zoo professionals, the "bite-sized" theory of eating live food states that, if it can be eaten in one bite, it is probably an acceptable spectacle for the public. If not, it would be too distasteful.

Traditionally, captive animals—and people, for that matter—were often simply an expression of the ability of certain individuals to go and bring them back, or have others do it. According to the French historian of the zoo, Gustave Loisel, collections of wild animals were first documented in places of worship in ancient Egypt, where the public could see lions being fed live prey at the Temple of the Sun in Metropolis and feed cakes and pieces of meat to the sacred crocodiles of Lake Moeris. The earliest recorded shows in Rome, given by Marcus Fulvius Nobilior in 186 B.C., used lions and leopards and a collection of animals given as gifts from the governors of Roman colonies and other foreign dignitaries. These also included hippopotamuses from the Nile, lions from Mesopotamia, and tigers from Persia (Ling Ling, the Panda at the National Zoo in Washington, D.C., was a diplomatic gift given in that tradition). All of the Roman emperors kept large collections of animals that were open to the public. Octavius Augustus (29–14 B.C.) had some 3,500 animals, which included 420 tigers, 260 lions, 1 rhinoceros, 36 crocodiles, and 600 African animals, including panthers, cheetahs, and 1 hippopotamus. Emperor Trajan (A.D. 98–117) had 11,000 wild and domestic animals. And, zoos are not only a Western tradition. In China, a park called Ling Yu—the "Park of Intelligence"—built about 1100 or 1000 B.C. for the Emperor Wen Wang, in the province of Ho Nan, walled in nine hundred acres and contained species of deer, goats, antelopes, birds, and an enormous quantity of fish. Meng Tseu, the Chinese author who wrote about the park, also told of the great park of the Emperor Chi Hang-ti, which contained birds, animals, fish, and three thousand types of trees and plants, and included replicas of all the palaces and castles he had destroyed during his campaigns. A better-known collection among Westerners was that of Kublai Khan. In the thirteenth century, Marco Polo reported that he had sixteen miles of parkland where game animals of all sorts were kept for hunting, including a collection of tigers, panthers, leopards, lynx, rhinoceroses, and elephants.

Some early collections were for science, not vanity, although one does not necessarily exclude the other. The collection of animals at Alexandria owned by Ptolemy II (283–246 B.C.), the largest of its kind in the Hellenic World, was established because Ptolemy had a particular interest in natural science. During his campaigns, Alexander the Great assembled a collection of animals for his tutor Aristotle, who used the collection for his works of natural history. Later, a great deal of scientific study was carried out on the collection of Louis XIV in France, who had some 222 different species, including an elephant given to him by the King of Portugal. At the time, it was judged to be the most popular animal in the collection. During the French Revolution in October 1789, an attempt was made to free those animals en masse. In the end only a few were freed, the dangerous ones were left in cages, others were sent to the slaughterhouse, while the remainder, presumably freer in spirit, became the collection at the National Museum in Paris, founded in 1793.

In the next century, zoos were established in Basel, London, Bremen, Philadelphia, and New York, among other places. Today, there are 156 accredited major zoos and aquariums in North America alone and another 600 or so that would qualify as someone's idea of a zoo. If, as a zoologist, Dr. Newman inherited a Western scientific tradition of investigation and classification, as an aquarium director he inherited another tradition: a uniquely human desire to hold in captivity large numbers of other creatures from all over the world. That tradition now includes two killer whales, Bjossa and Finna, "ambassadors of their species," as Dr. Newman likes to call them. Those two animals, symbolic of the Northwest Coast of America, were pulled from the sea near Iceland in December 1980. They were packed in crates, with their tails and dorsal fins covered in ice, and flown fourteen hours to Vancouver, joining forty-two other killer whales in captivity around the world.

Students enter John Ford's office and announce they want to study whale and dolphin language in the same way they might announce they want to study Spanish. Ford, the curator of marine mammals at the aquarium, in his mid-thirties, red-haired with a lot of freckles, explains to them that in fifteen years of studying killer whale communication, which includes a Ph.D. from the University of British Columbia, he has not yet reached the stage of learning their language. He tells them his approach is "more classical"—to correlate the activities of the whales with the sounds they make—and if their desire is to speak to the whales, they might find his work rather boring. Most of the potential linguists, though, have seen the spectrographs of killer whale sounds, so they'll say something like, "They look so complex, they must *mean* something." Ford responds by showing them another spectrograph, which looks similar and equally complex. They nod seriously. Then he tells them it's a tree frog.

As I entered Ford's office through a door in the wall of the underwater viewing area, I felt as if I were passing through a doorway into a place where all the secrets were kept. Behind me, the children in the underwater viewing area were clambering over a fiberglass model of a killer whale dorsal fin; ahead of me, a silent room was illuminated by the pale green screen of a Macintosh computer and the blue-green cast from two windows on the killer whale pool. As I walked in, Bjossa, Finna, and Whitewings floated by, arranged in a vertical hierarchy, of sorts, resting after feeding.

Ford was at his desk. Near his hand were two stacks of pink telephone messages, each about four inches high. The telephone would ring constantly while I was sitting there. Beside the stack of messages was a photograph of his wife, taken in the Azores, wearing a Panama hat and surrounded by large white camellias. They spent their recent honeymoon in Patagonia, recording and observing killer whales, thus making even his honeymoon part of his ongoing study. What started in 1978 as a two-year graduate student project now will probably continue for the rest of his working life.

Killer whales have good eyesight, but rely on underwater sound for navigation and communication. They force air through elaborate structures in the nasal passages beneath the blowhole to generate ultrasonic clicks less than a millisecond in duration. Using a fatty melon in the forehead, these clicks are focused into a directional beam and bounced off objects—a process called echolocation—which enables the whales to form an image of their surroundings. They also produce canarylike whistles more than ten seconds long, and loud complex calls that can be heard for miles underwater. As might be expected, the whale can hear with exceptional sensitivity, mainly using its lower jaw, rather than its ear canal, which is closed. A fat body in the jawbone carries sound to the middle and inner ears.

Ford's interest was spurred by the idea that such acoustics enhanced a social system among the some 350 wild whales that inhabit the Northwest Coast. That notion, born of his own naivete, "was not encouraged," he told me. Studies by Michael Bigg, at the Pacific Biological Station in Nanaimo, British Columbia, Kenneth Balcomb in the San Juan Islands, and others had already identified northern and southern "resident" communities, along with other groups—"transients"—that roamed freely up and down the coast. Furthermore, they had discovered that these communities contained family "pods." But, dialects are rare among animals, and Ford said if he'd had more training in mammalian communication, he probably wouldn't have started with a hypothesis that marine mammals had pod-specific dialects.

Over a period of years, with support from the University of British Columbia, the Canadian government, the aquarium, and some cheap Navy surplus hydrophones—underwater microphones—Ford proceeded. Eventually, he found that each resident pod could be identified by sound alone and had a set of about a dozen different types of "discrete calls," a two second or so pulse (several thousand pulses per second) producing high-pitched squeals and screams. In other words, each pod had its own "dialect." He also found that the pods could be grouped into four "clans," pods that are acoustically similar or share a traditional set of calls.

As Ford explained his findings, Bjossa, Finna, and Whitewings floated by and Ford pointed to them. "After they feed here, they arrange themselves in a dominance hierarchy," he said. "They probably do that in the wild. We think they do. But we can't see them underwater when they rest after feeding in the wild and that's one of the problems we have. For example, the first calf here died of starvation. We know that almost half of all the calves born in the wild die. Do they die of starvation? Is that typical of first-time mothers? Was the death here typical, then? You see, we're never quite sure how much of the behavior *here* is duplicated in the wild. Basically, in the wild, much of what they do is not seen. We can record sounds, but it's difficult to tell what they're doing. So, for the dialects, our level of knowledge is reasonable, but we really don't know what those dialects mean.

"That's one of the reasons I'm hoping for a live birth, so we can look at vocal development in the young," he added. "We can begin to make behavior-to-sound correlations, since it will be possible to look at behavior in some detail. Of course, the range of social interaction is limited in captivity, so that's a drawback."

Finna glided up to the window and put his head against the glass. Was he looking in? Attracted by the greenish glow? He sat there, his right eye staring in at us. Finna is multilingual, Ford said. He "speaks" the dialect of his original Ice-landic pod, the northern Vancouver Island dialect he learned from the famous Hyak, and the southern Vancouver Island dialect Hyak had learned from Skana. Killer whales are part of a select group of mammals capable of learning to repro-duce the sounds that they hear. Such learning was once thought to be exclusive to birds and humans. Now, it's known to exist in a few primate species, some seals, and in the dolphin family. The calls of the killer whales are most likely learned, rather than genetically determined, and young whales probably learn by mimicking their mother.

The multilingual Finna stared in at us, or so I thought. On the other side of the door, I could hear children stampeding down the ramp. I had read an article of Ford's that said when the whales are physically interacting, chasing, nipping, push-ing, and so forth, the kind of thing we'd call play in humans, they emit calls that are seldom, if ever, repeated in the same form. I had been thinking about the "seldom, if ever, re-peated" and I wondered if it could be construed as creativity. I imagined a sound masterpiece, a killer whale Miles Davis, going where no one had ever gone before.

"Do you think the whales are being creative when they make up a new call?" I asked.

He paused.

"They learn sounds," Ford explained. "They learn dialects. Those are facts. Or 95 percent facts. But, creative? No."

He went on to say that calling will often be contagious. One whale makes a call and it triggers the same call in oth-ers, and if they're socializing intensely you can get almost a "constant jabber," a situation that could provide the raw ma-terial for such new calls. He described it "like kids in a play-ground, a sort of emotional noise."

"I'm sure that's what these guys are doing. But, until we can see it, we can only say, this occurs, then this occurs. . . . It's frustrating. That's why I was happy to focus on dialects. It's very concrete."

Ford insisted his work was not an attempt at interspecies communication, unlike the experiments being conducted at the University of Hawaii by psychologist Louis Herman, where dolphins have learned to mimic electronically synthesized sounds broadcast into their pool by underwater loudspeakers. One dolphin was trained to mimic nine different manmade sounds upon command, then to associate a specific sound with an object; in effect, to label it and repeat it. Despite this success, the experimenters said that it was "foolhardy" to conclude from the experiments that dolphins were particularly close to humans in what is called "linguistic ability."

Of course, such a conclusion reveals a presumption that, if another creature is truly intelligent, it will be reflected in a linguistic ability similar to that of humans. It's a determination of intelligence that is defined by the terms of our syntax. If we can't fit the sounds made by another animal into the order or structure we have developed, we don't credit them with language, and it's a short step from there to not giving them credit for intelligence or even conscious thought. But, if languages exist for making ourselves understood, for creating and strengthening social bonds, then words are not the only important thing. How one speaks matters. We gesture, move our eyes, raise eyebrows, use our hands; all of that is part of the delivery, even if the words themselves are not understood. So, while languages are quite different, among humans, let alone between species, what they do is not so different. Why should the form of the language matter so much? Yet, the form of language is considered crucial by many biologists. Heini Hediger, author of the "zoo bible," visited W. B. Lemmon and Roger Fouts in Norman, Oklahoma, to observe their use of American Sign Language with their chimpanzee, Washoe, and later he visited Francine Paterson of Stanford University and her gorilla, Koko. In neither case, he wrote, was he convinced that the experiment "demonstrated actual communication by means of a formal language." The key words there are formal language. He explained his objections at a conference organized by the New York Academy of Sciences entitled, "The Clever Hans Phenomenon: Communication with Horses, Whales, Apes, and People." The phrase "clever Hans" referred to a celebrated horse, Hans, who before the First World War in Berlin was supposedly able to think like a human, do sums, speak, and so forth, although it turned out that Hans, a student of body language, like the killer whales, was relying on simple signals that had gone unnoticed.

In other words, signals are not speech, which is true, although the trainers in the pool at the aquarium used them each day to great effect to communicate across the species barrier. Is it only speech—and our narrow definition of speech—that qualifies as intelligent communication? Or is the speech barrier something else again? What, exactly, would an animal have to demonstrate before satisfying our criteria for intelligence? What hurdle would they have to leap? If we gathered enough information to indicate that they were communicating intelligently, what then? Would that affect the way we allow ourselves to treat them?

For example, in a field study done by Tom Struhsaker of

the New York Zoological Society, vervet monkeys in the Amboseli National Park, Kenya, were found to have different alarm calls for their three main predators—leopards, eagles, and snakes. Each call seemed to elicit a defensive response appropriate to the predator. If the leopard call, a barking, was given, the monkeys would run for the nearest tree. If it was a cough, for an eagle, they would look at the sky and head for dense scrub. A chuttering, for a snake, would put everyone on their hind legs looking at the tall grass. Later, this study was confirmed by Robert Seyfarth and Dorothy Cheney, who played taped alarm calls varied in loudness and duration from loudspeakers hidden in bushes. The monkeys reacted to recordings made to sound more or less urgent and even with no predator in sight. Evidently, what drove the response was not the sight of the predator or urgency of the call, but the meaning it carried. Furthermore, they found evidence that vervet monkeys must learn the appropriate use of the various alarm calls. At first, infants tend to give any of the three alarm calls to all of the predators. With time, however, they learn to associate each alarm call with a specific predator. The research further demonstrated that monkeys can not only identify other individuals by voice alone but they are also aware of the relationships around those individuals. For example, when they played back the screams of an infant, there was of course a strong response from the mother, who looked toward the calls of distress. The other monkeys, however, did not look in the direction of the screams. They looked toward the mother, apparently recognizing in the screams not just distress, but whose infant was in distress.

Vervet monkeys seem to learn, then, as do birds, killer whales, and humans. Killer whales have dialects, as do humans and birds. The song of a bird develops in stages, not

unlike that of speech in a human child, which begins in con-
fusion and goes on by means of experimentation and learning. In the beginning, a child produces many kinds of sounds
that it will later fail to use. Likewise, a bird may at first produce four or five times as many kinds of syllables as it will actually use when it reaches the final stage of development of
its song. As one researcher described it, some stages might be
considered "vocal play . . . a search for inventiveness—nov-
elty for its own sake." Does the same sort of progression take
place with killer whales? Ford was eagerly watching Bjossa's
pregnancy in the hope of being able to investigate that possibility. How many calls might it develop before settling on its
mother's? How many mistakes will it make? How many new
things might he or she try? Will it sound like the "constant
jabber" of children in a playground? As I sat with Ford, I
could hear children on the other side of the door, like all animals, screaming, laughing, using speech and anything else
they could think of to communicate or fail to communicate.
Our benchmark, our hurdle, is formal language or linguistic
ability, but speech, after all, may not be the pinnacle of communication. Make a face. Touch my arm.

"If vervet monkeys can do it,
there's no reason killer whales can't do it. I mean, not only
recognize other animals, but recognize relationships," said
Ford. "We can do it. When we pick up the phone, everyone
says, 'hello.' We use the same word, but you recognize who it
is by the voice. It would be nice to show that someday in
killer whales.

"Today, though, what's got me buffaloed is this," he went
on. "You take a whale out of his or her group and he makes

all the same calls as his group makes in the wild. So, what does the call mean? I'm beginning to think these calls are not telling us about behavior. I've started thinking that the call types should be looked at more as 'carriers' and that they don't say so much about the environment. The way they make the calls has to carry the meaning. So, let's say the call means, I am from this pod—in other words, the call is an identification for the pod and the individual. The variations in the call could convey other important information, such as emotional state." That has already been found in bottlenose dolphins, which have signature whistles—a carrier—and the modifications of that whistle carry emotional behavior information.

The telephone rang. Ford ignored it, rolled his eyes, got up, and sat in front of the Macintosh computer. He retrieved a file of identification symbols and scanned it for the male/ female symbols, so he could continue updating lineages in a handbook of the killer whales on the British Columbia coast. He's now trying to work closely with several people from Norway, Argentina, and the Antarctic, so they can begin comparing the whales from those regions. "It's just fascinating that you could have this mosaic of social traditions in socially insular populations," he said. He explained that killer whales in the Arctic apparently don't like ice. They're only seen in the high Arctic in August and September, when the ice is gone. Yet, in the Antarctic, they routinely follow the first ice breaker to look for shore seals, and they seem very comfortable with the ice. In Argentina, they beach in order to hunt seals, a highly specialized, highly risky behavior. "Of course, they've had millions of years to develop in a completely different environment," he said. "We should expect them to surprise us."

The aquarium is now beginning the first of several areas of research that cannot be done in the wild. For example, they are setting up an experiment to test how well the whales can hear, which they hope will have practical application to problems with the black cod fishery in Prince William Sound, where commercial fishermen are upset that killer whales are taking fish off their lines. It's hoped high-frequency sound could perhaps drive them away. In addition, Ford is preparing to establish a monitoring spot in the Queen Charlotte Islands, four hundred miles northwest of Vancouver, to expand his ongoing study and tie in with the proposed Pacific Canada exhibit. The focal point of the exhibit will be a two-story-high tank that will reproduce a scene from the South Moresby area of the Queen Charlotte Islands, a national marine park reserve, where several groups of killer whales had been spotted that had never been seen before. They were being called the "mystery" whales.

That, Ford argues, is why whales should be in captivity, not to jump through a hoop to make some money off them, but for research and conservation, and the aquarium must be very clear about that mission. "It's intense, because people feel a special relationship with them," he said. "I don't think that feeling would exist if they weren't here at the aquarium. It's not a circus act. My feeling is that it establishes a bond with the animal and the impact is in the interaction between people and animals. That live presence is very, very powerful, especially for kids." Still, he added, even with a bond established, they are not people in a tank, imprisoned, as some critics of the aquarium have suggested, but "alien beings, so different from us they might as well be from another planet."

"Why do people feel a special relationship to the whales?" I asked. "Why do they attract so much attention?" Finna bumped against the glass. Whitewings dashed by, just beyond him.

"Because they encapsulate so many things," he said. He ticked off the reasons. The killer whale, symbolic of the British Columbia coastal wilderness, is dramatic, striking, not numerous, and widely and easily recognized, since they're promoted on television, greeting cards, and posters. "There's a lot of hype," he said.

It was a Friday, in the early morning. I had been at the aquarium about two weeks, and I was sitting on a small hill behind the sandstone backdrop, at a spot near the fir trees and a small Japanese garden. Below me, at the foot of the hill, a blackbird was drinking from an artificial waterfall. I could see into the habitat. A maintenance worker was hosing down the concrete seats. The whales were drifting around the pool, perhaps napping, regularly tossing up a "kwhoof." Doug Pemberton came out of the cave and crouched by the side of the pool. Finna quickly poked his head out of the water at Doug's feet. Doug fed him vitamins, and as he did, the heron—curious, it could be food, after all—launched himself from the fir tree, made a long, swinging approach, and settled on the concrete a few feet from Doug. He kept his distance, but cast longing glances at the feeding. Off to my right, the metal gate on a concession stand was lifted, screeching on rusted metal hinges, and children began to circle the stand, wondering what they had to do to get some sort of treat. The woman who had opened the stand gestured with her head and the kids formed themselves in a line.

It looked and felt peaceful. A few days before, though, I'd

been given a quick underground tour by John Rawle, a husky man who wrote reminders to himself in blue ink on his right hand. If Dr. Newman was the captain of the ship, Rawle was the growling first mate who kept the whole thing running. He called the killer whale habitat a combination swimming pool and life-support system. "They have to live in it all the time," he said of the whales. "They can't get out for a shower." The water itself is natural sea water, kept at nine degrees centigrade in the winter and sixteen degrees centigrade in the summer, with a turnover of a million gallons every 150 minutes. He showed me the original mechanical area for the aquarium, which was a room about thirty by fifty feet. Today, it is about twenty-five thousand square feet, and a tank for sharks now sits over the original mechanical area for the entire aquarium. Each of four vacuum filters, which cost $60,000, but would cost $125,000 to replace, covers about thirty feet and filters 1,700 gallons of water per minute, using a forty-horsepower motor and two fifty-pound bags of Witco filtering powder per day. The powder, which is also used for filtering wine and scotch, is layered onto fiberglass plates and serves the same purpose as the bag in a vacuum cleaner. The collected waste material is taken to a sewage disposal plant; 160 tons a year at a cost of four thousand dollars per month.

Rawle peered through the glass at the end of the run through and said, "The water is nice this morning. I wish Dr. Newman were here to see it." The water will always be slightly hazy from algae and "particulate matter," he explained—killer whales have practically no bladder and continually urinate—but the clarity of the water has nothing to do with its purity. Murky water can still be bacteria free—and safe—which is accomplished by first treating the water with ozone, then chlorine bleach. The ozone is removed by using biological filters. Certain strains of algae, however, eventually build up resistance to low residual chlorine concentrations, so periodic "shock treatment" is needed.

As I sat there that morning, feeling the breath of the breeze, watching Doug Pemberton carry on some kind of interspecies communication and fend off the entreaties of the great blue heron, I knew that if I entered the cave and went below ground I'd find John Rawle racing around among dozens of basement rooms, wrench in one hand, putty in the other, sweating profusely as he plugged leaks in the concrete and made sure the salt water hadn't corroded everything. "Insidious," he had told me. "It eats things up." Chemical reactions would be exploding like fireworks around him and the whole machine would be sucking and pumping to keep alive two killer whales and their high-speed friend. It was as if I were looking at a whaling ship, only the prey was alive and swimming around in the hold.

The bronze killer whale that sits outside of the main entrance to the aquarium is eighteen feet high and weighs more than one and a half tons. It stands on its tail flukes, with its dorsal fin thrust toward the sky. Northwest Coast artist Bill Reid, who is part Haida Indian, designed it with mountain trout in the tail flukes, a wide human face carved in each side, and a human face in the blowhole. I imagined it to be Dr. Newman's face, a professional life inextricably melded with the killer whale. The sculpture was explosive, uncluttered, and I didn't like it, for the very reasons some would find it attractive. The wood sculpture I'd come to associate with the Haida invited intimacy. The huge

bronze repelled. Wood is warm. Bronze is cold. It will probably last hundreds of years longer than a similar sculpture in
wood.

There was something else I didn't like about the sculpture.
Its base was a reflecting pool, which looked as if it had come
from another age or another mind. The pool was about
twenty feet in diameter and water flowed from the middle,
but so smoothly the movement wasn't visible. Although the
sound of rushing water could be heard as it flowed over the
outer edge, the water looked as still as glass. The effect pro-
voked curiosity. What was that sound? Was that water? For
the answer, touch the water flowing over the pool's outer
edge. Eureka!

After examining the phenomenon for myself, I wondered,
how in the hell did this pool get here? The best way to find
out would be to talk to Bill Reid, although I knew that
wouldn't be easy. He's one of the world's great artists, one of a
handful of craftspersons credited with keeping the art of the
Northwest Coast Native Americans alive, praised by Lévi-
Strauss as an "incomparable artist" who "tended and revived"
the flame of the art that was close to dying. He was also sev-
enty-one years old and not in the best of health. I was told
he had a studio on Granville Island, a refurbished upscale
shopping area on an island in English Bay, which was reached
by bridge from downtown. It wasn't far from where I'd taken a
dunking in the bay's fifty-degree water, while kayaking a few
days before. We had been practicing rescues. After my in-
structor from the EcoMarine Kayak Center had flipped my
kayak, he decided to depart from the accepted script. Instead
of helping me, he flipped his kayak and started flailing
around in the water, acting as if he was drowning. I suppose
he thought he was making the session more realistic and in-

teresting. He was mistaken. "Now what do we do?" he screamed. "Call the Coast Guard," I screamed back, which was not the correct answer, but which I still think was good advice.

In any case, I called Reid at his studio. He put me off. I called him back each day for several days. Finally, he agreed to talk with a gruff, "Okay, let's get it over with." I rushed over. He was standing outside when I got there, leaning on a cane and wearing blue-striped pajama bottoms. His long, fine, snow white hair was swept back and glittering in the afternoon sun.

Reid's mother was Haida. His grandmother was a native of Tanu, a village on South Moresby in the Queen Charlotte Islands, which was abandoned before the turn of the century, part of a gradual retreat northward for the Haida population in the face of smallpox. In the village of Skidegate, she married Reid's grandfather, Charles Gladstone, and their union eventually made Reid a great-nephew to one of the master Haida carvers, Charles Edenshaw, whose drawings can be found in Franz Boas's book *Primitive Art*.

Reid was born in 1920, the year Edenshaw died. He didn't see Edenshaw's work until 1954, when he was attending his grandfather's funeral in Skidegate. By then, Reid's native roots had withered. He was working as an announcer for the Canadian Broadcasting Company. While at the funeral, though, a friend and carving companion of his grandfather's encouraged Reid to go and look at two of Edenshaw's gold bracelets, because they were, Reid was told, "deeply carved." He did, and since then has often been quoted as saying, "The world was not the same after that."

It certainly wasn't, for Reid or anyone else associated with native Northwest Coast art. At about the same time, the art began to be "rediscovered" and Reid played a significant part in that rediscovery. He became part of a team from the British Columbia Provincial Museum that went north in 1955 and 1957 to salvage totem poles from abandoned village sites in the Queen Charlotte Islands. He met Mungo Martin in 1957, a Kwakiutl Indian who had spent his life keeping Northwest Coast artistic traditions alive, and spent his vacation from the CBC working with him on a totem pole, which now stands at the Peace Arch on the Canadian-American border at Blaine, Washington. Reid found wood carving came to him quite naturally. Between 1958 and 1962 Reid and another Indian carver, Doug Crammer, built two traditional houses and carved seven poles and other massive wood pieces, which are now the grounds outside the Museum of Anthropology at the University of British Columbia. By 1963, Reid, having begun to study the art for its artistic principles, left the CBC and established himself as a full-time artist.

There were several benchmarks in his forty years of work. In 1978, he carved a fifty-foot pole in his mother's village and donated it to be raised in front of the new Band Council House. Carved in cedar, Raven and the First Men is the creation myth of the Haida and is displayed at the University of British Columbia Museum of Anthropology. It is, perhaps, the single work by which Reid is most widely known. In 1986, Reid carved a canoe for the Canadian Expo. And, in Washington, D.C., a city of monumental sculpture, a Reid bronze called the Spirit Canoe stands in front of the Canadian Embassy. At the same time, he was encouraging other, younger native artists, such as Robert Davidson, who lived and worked with Reid in the mid-1960s and today is as well known and respected as Reid himself. Through it all, Reid kept his modesty by remembering the phrase "deeply carved." It meant simply that something must be "well made," which took time and attention to detail, so each object could become, in his words, "a frozen universe."

I followed Bill into his studio. Was I really going to tell him I didn't like his killer whale? We passed through a small reception area, a desk piled high with papers and photographs, and into the spacious studio, which was wonderfully chaotic and filled with the sweet, warm smell of cedar. When Bill sat down, he immediately picked up two cedar slats from a large pile at his feet and began trying to fit them together. A bulletin board for his daughter, he explained. His hands were smooth, pale, delicate, and mottled with age spots, and when he stopped pressing the wood together, they began to shake uncontrollably. He had Parkinson's disease. He pressed the wood together. The shaking stopped. The wood was bulletin board and remedy.

We sat in two chairs in the middle of the studio, as if a space had been cleared for visitors amidst the chaos, his a cane chair, mine an overstuffed ancient leather, which promptly tipped over. While I righted the chair, Bill continued to press the wood together, and I wondered if the chair was his way of keeping visitors slightly off balance. If so, he had succeeded. I was still trying to think of a way to phrase my questions. I thought I should avoid starting with, "Well, Bill, that's one ugly killer whale. What happened?"

"I wanted to ask you about the killer whale bronze," I said. He nodded, rose, and indicated with a wave of his hand that I should be patient. He shuffled away, in search of another piece of wood. I recognized that shuffle. I had seen it in my father, who died of Alzheimer's disease and whose speech, sense of touch, and ability to walk had faded and fallen away, one by one, like the petals from a wilting rose. From the rear, Bill looked remarkably like my father as he had approached his end, white hair swept back and forever shuffling until he could shuffle no more. Reid returned. He nodded that I should begin again.

"Well, to begin with, where did that abstract base come from?" I asked.

He spoke slowly. His voice was low pitched, almost inaudible, when he began, although it gained strength with each word. "Well, Murray Newman approached Liz Nichol—she represented me at that time—and asked if I was interested in doing a large piece. Originally, I imagined different materials and below the whale it was supposed to look wild, like Hecate Strait. That's between the Queen Charlottes and the mainland. But then Mary Margaret Young, she saw a pool in San Francisco or someplace and she wanted one and because she has all the money in the world she got it." I was later told that Mary Margaret Young was the daughter of Earl Finning, who owned Caterpillar Tractors, and she donated the money for the pool.

I paused. "Well, do you like it?" I asked.

"I hate it," he said loudly and fully in control. He grinned. "Between that pool and the peacock that's always flying around there, yelling its head off, the whale doesn't stand a chance. Nobody looks at it."

His legs were shaking. He shifted in his seat, but he was still smiling, pleased with his answer.

"The original idea was a sort of lord of the underworld, a five-finned killer whale, and it was to be on top of an island," he said. "So, after the design was finished, I looked at what it was going to take to get it done and I said, 'I'm too old and too fragile and my life is too short,' so I backed out. At about that time, Robert Davidson came by. He was busy playing the Haida game at that time, dressing in Haida clothing and so forth, probably eating Haida food, I don't know. In any case, he decided he had to resurrect his career, so he phoned and said he was looking for a nice big job. I said, you're just in time, because I have a nice big one I don't want. Call 'em up. I told him what Murray wanted and told him, give him a nice whale and he'll be happy. So, he came up with the idea of two whales becoming four whales." He grinned. "They turned it down."

He paused. The wood slipped from his hands into the pile at his feet. His hands shook as he reached to retrieve it.

"So, they came back to me and said, couldn't it just be a whale? I had a little carving I'd done in the meantime, for no particular reason, and I showed them that and they eventually drew it up. And there it is, disappearing into a reflecting pool."

"How do people respond to it, do you think?"

He waved his hands around at the studio, which had a thirty-foot ceiling with wide beams. "I actually rented this place for that job, and close to the end, we had it all assembled, made of plaster—a bronze is carved in plaster, then cast, you see—right under the skylight here, and kids would come in and say, 'Look at the totem pole.' So, they knew. But now, it's something else. I don't know what it is. I suppose, though, at the very least, it introduces people to the art, which they seem to understand and communicate with. I know this—we can't expect to feel the impact of art as long as we keep it in little glass cages."

He seemed to be rolling. I didn't have to ask any questions. He was enjoying himself.

"As for the notion of the Haida art," he said, exchanging his cedar slats for others, pressing them firmly together, "the Haida have been told so often that they're great artists by virtue of being born Haida, that they no longer think you have to work at being an artist or that art comes from application or from something that's well done, well carved. They think you can just do whatever it is you want to do and, because you're Haida, it will be art. Like all you need is a sign on the door. No training. No skills. Nothing. It's just, the fishing isn't good, so I'll turn to silver carving."

He smiled. It was a sly smile, though.

"That isn't true of everybody," he added.

He got up slowly and shuffled to the rear of the studio, where a twenty-seven-foot canoe lay. I followed. He pulled some slats from inside it and examined them.

"Did you see the inscription on the killer whale?" he asked. "It says, 'Above everything is God, above God is the killer whale.' It's a nice thought."

He paused, examining his slats and sliding them together.

I told him I might go to the Queen Charlottes. He nodded slowly. "There are fifty supernatural killer whales up there," he said. "If you see one, let me know." He paused. "That's how it will end for me, with the fishes in the water up there. Maybe one of these killer whales will get me. We all end up eaten by worms anyway. It might as well be a killer whale."

"I was told they don't eat humans," I said. "They might just ignore you. They wouldn't know they were supposed to eat you."

He thought about that for a second, smiled, and said in almost a whisper, "Well, that doesn't have a very mythic ring to it, does it?"

After that, he seemed to have suddenly run out of energy. He nodded for me to follow, pressed his wood together, and shuffled out to the small reception area. He started pawing through the big stack of photographs. Pictures of the "Spirit Canoe" were on the wall, and he showed some others from the pile—pictures of sculpture, of canoes, pictures of Bill and Lévi-Strauss and Bill's wife Martine, twenty-five years his junior, who was once a student of Lévi-Strauss and now a Ph.D. in anthropology. None of them seemed to satisfy him. He nodded again and I followed him, shuffling, shuffling, back into the studio. He pushed some wood aside. There was a plaster cast of the Raven and the First Men sculpture on a table. He seemed very tired now. He put a small, wooden carving out for me to see. It couldn't have been more than four inches high, wooden, intimately detailed, a killer whale with its dorsal fin in the air, a human face in its blowhole, and its tail flukes resting on a stormy sea.

He let me look at it for a few minutes, then his assistant came in and said it was time to go. "The Doctor will be upset if we're late," she said, referring to his wife. Bill said a meeting had been set up at his house with two lawyers. Some people were unhappy with the way his work was being marketed and the two lawyers were supposed to take over the show. "Might be worth seeing," he said, and invited me to come along. But he sounded tired and unenthusiastic and I felt I had already imposed enough. I also had a vision of two lawyers picking over his work while he shuffled around the house from piece to piece. I didn't want to see it.

Instead, his assistant drove him home, and I took a half-hour city bus ride out to the University of British Columbia to see the Raven and the First Men sculpture. I entered the Museum of Anthropology, designed by Canadian architect Robert Venturi, and walked down an inclined ramp past grave figures, totem poles used as house posts, and wooden sea lions holding a huge cross beam up with their noses. I turned right, passed the research collection, and there, on the left, was Bill's sculpture, standing by itself in a circular room with a circular pedestal and a bed of sand, surrounded by rust brown velvet rope, the kind used to keep people out of swank clubs. A circular skylight was directly over the pedestal and three track lights illuminated the sculpture.

I walked up the ramp and around the sculpture. About four feet high, it seemed to throb and twist on its base, as if the Raven, squatting on the clamshell and gripping it with its talons, was about to jerk the shell out of the sand toward the sky. Large men were forcing the clamshell open and climbing out, men with open, expressionless faces, butts in the air, their testicles in my face. The beak of the Raven was thrust forward. Its wings draped to the sides and a human face had been carved in its tail. In Bill's retelling of this Haida creation myth, the Raven, a trickster full of curiosity and with an itch to provoke things, who has always existed and will always exist, found the enormous clamshell on the beach. He coaxed the men from the shell, and the men—the first humans—found themselves on the beach, "confused by a rush of new emotions and sensations." They "shuffled and squirmed, uncertain whether it was pleasure or pain they were experiencing."

I sat and drew it. I imagined the adz on the wood. The smell of Bill's studio filled my nose. Then I sat back and dozed. When I woke up, I immediately rose, walked over, reached out over the velvet ropes, and touched the left wing tip. I thought, better than the killer whale bronze, no distractions, no pool of water, and more intimate. Bill had said it himself, though—if art was going to affect people, it had to come out of the glass case. Here, all the sensuality, all the sloppiness and juice of life was behind ropes, on display, like two killer whales in a tank.

San

Juan

Island

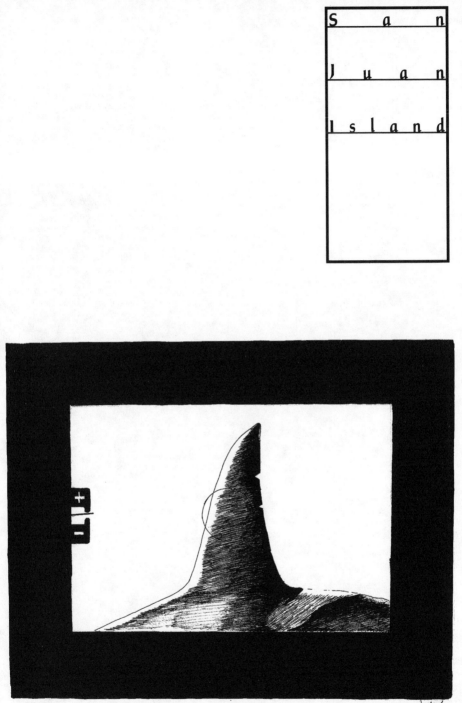

SAN JUAN ISLAND

The Center for Whale Research is a grand title for Ken Balcomb, a band of volunteers, and assorted machinery, not all of it in working order. Two ferry-boat rides—Vancouver to Sidney, British Columbia, then Sydney to Friday Harbor in Washington State's San Juan Islands—brought me there. Even though they are only fifty miles from Vancouver, I felt as if I'd passed into another climatic zone. Normally, the Pacific Coast receives more than 150 inches of rain, but it was sunny and hot the afternoon I arrived. The San Juans are in a protected rain shadow, which also takes in Victoria, on Vancouver Island, and the northeast tip of the Olympic Peninsula. I might as well have gone to Southern California. I had also passed into a psychic zone where the whales were no longer known as killers, but orcas, from the Latin *Orcinus orca*. Once I'd left the aquarium, my first use of the term "killer" whale was met with embarrassed silence and those around me looked at each other, wondering, doesn't he know any better? I wasn't corrected for the gaffe, but the word *orca* was stressed in every sentence thereafter until the message sunk in, much like intensive foreign language instruction.

Friday Harbor, population 1,500, is the largest town in the San Juan archipelago, which is made up of, depending upon the tide, anywhere from 450 to 750 islands, most of them tiny, uninhabited granite and sandstone clusters. Only four islands, with fewer than 10,000 people, are served by the ferry system: Lopez (population 1,200, mostly farmers), Shaw (population 130), Orcas (the largest and most developed, with some twenty resorts), and San Juan Island, on the east side of which the original natives, the Lummi, believed that human life had begun in a wilderness Garden of Eden.

The paradise was plowed under when Spanish explorers arrived in the 1770s and 1780s and the usual business followed, with English explorers, Russian explorers, claims, investigations of claims, George Vancouver's 1792 expedition, which, in turn, brought the Americans, joint occupancy, and finally, in 1846, a settlement—the Treaty of Washington—that gave everything above the Forty-ninth Parallel to Great Britain, but left ownership of the San Juan Islands murky. That situation didn't seem to bother anybody until 1859, when a pig owned by a British settler wandered into the potato patch of an American farmer, Lyman Cutler, who—typical American of his time—promptly shot the pig. United States Army captain William S. Harney quickly dispatched a platoon of troops to back up Cutler. The British responded with their own troops. Before another pig could snuffle another potato, more than a thousand troops were on the island, nine miles apart, accompanied by three British Royal Navy ships and one American ship equipped with a howitzer.

It had all the makings of the 1982 Falkland Islands war, but incredibly, not a shot was fired. The troops stayed for thirteen years and the stand-off became known as the "Pig War." In the interim, the Americans fought a Civil War. The troops on San Juan Island were able to sit that one out, while 620,000 of their fellow soldiers died, and it's possible the carnage dampened their enthusiasm for bloodshed. Instead, the two warring parties eventually held holiday banquets together. The dispute was finally settled in 1872 by arbitrator Kaiser Wilhelm, who awarded the islands to the Americans. The British departed, leaving behind a rectangular garden, where they had grown flowers. A few miles up the road from the garden is Ken Balcomb's place.

Ken picked me up dockside in a faded turquoise GMC Custom Truck. In his late forties, he was wearing sandals, worn jeans, and a T-shirt that once had a logo. His shoulders were rounded, his chest sunken, and he had the hips of a twelve-year-old. His curly brown hair and full beard were flecked with gray and his nose was sunburned a painful shade of purple. After making a quick stop for a hundred dollars worth of groceries to feed the volunteers, he followed a winding, two-lane road to the west side of the island, thirty minutes away. In the truck, he told me that when he first moved to the San Juans, in 1976, visitors had to hunt to find his house. Now, there are condominiums in Friday Harbor and he has an address—1359 Smugglers Cove.

The site sloped gently down from road to shore. At the top of the slope, where Ken's house sat, the sun was hot and the air was still. He lives in a partially refurbished barn nick-named "The Shop" because of all the junk inside. The inte-rior decoration was provided courtesy of the "Polyxeni," a freighter that wrecked near Silver Bank in the Bahamas. Its five signal flags were draped inside and the ship's registry was tacked near the front door. Nearby were two outhouses, one of which had a beautiful view of Haro Strait and was equipped with toilet paper from Envision—"Environmentally friendly paper products." A car port was attached to the house, where a 1934 Deluxe four-door Ford V8 was hidden beneath a baby-blue plastic tarpaulin. Ken purchased and re-stored it because it was the same car he had in high school. Later that evening, he would race from one end of the prop-erty to the other on a hundred-yard test drive.

We walked down the slope past a navy blue pickup truck, a gray Datsun, and a beige Lincoln Continental, which were lined up behind a log, as if waiting for a "Prices Slashed" sticker. Nearby, there were a dozen gnarled apple trees and a garden. At the bottom of the slope, a chilled breeze rose from the water. The waterside split-level house, used by his volun-teers, had a front deck and an extension on which someone had painted an orca mother and her calf. It looked like the prow of a ship. From that spot, the Haro Strait stretched 180 degrees, sparkling in the sun, with the thin outline of Van-couver Island and the jagged, pale blue of the snow-capped Olympic Mountains on the horizon. Two other official vehi-cles, a red Chevrolet Suburban and a buff GMC Sierra 4 x 4, were parked near the split-level. I was told that all the official vehicles were necessary, since it's hit and miss which one might be working at any one time.

Anchored about fifty yards offshore were a twenty-foot Boston Whaler and a thirty-seven-foot baby-blue and white Trimaran, a motorized sailboat. The Trimaran was donated several years ago by Prentice Bloedel, a grandson of the Bloedel family of MacMillan-Bloedel, the largest timber company in Canada. Prentice is considered the black sheep of the family because of his long interest in conservation, and more than a decade ago he helped Ken buy the property for his research center. On shore, another boat was laid up on blocks, for the time being. Paul Whittier, a local philanthropist, built it to look like a 1920s Long Island shore boat, then donated it to the center.

When we arrived, Ken surveyed the surroundings and declared that he was a "slave to all these machines." It occurred to me that a man who studies orcas throughout the summer, humpback whales throughout the winter in the Dominican Republic, "patching it together" for sixteen years and believing he'll do it the rest of his life, is more likely a slave to something else. "Lots of people couldn't stand the life I lead," he said, after the quick tour of all he owns. "And I'm glad they can't."

Orcas can be found in every ocean of the world, the largest concentrations being in the colder seas, especially off Japan, Iceland, Norway, Antarctica, and the Pacific Northwest, from Washington State to Alaska. There is no reliable estimate of world numbers, but they do not appear abundant. The population in the Pacific Northwest is perhaps the densest and certainly the most studied.

Ken came to San Juan Island because, he says, "this is where the whales are," and because the National Marine Fisheries service was embarrassed that, if it wanted to know how many orcas it had in United States waters, it had to ask Canadians. Ken, fresh from eight years in the navy as a pilot and an oceanographer, was awarded a contract to take a census of the whales in the Puget Sound, to verify the work of Michael Bigg, the Canadian marine biologist. Bigg had developed a means of identifying killer whales by photographing the dorsal fin area and examining the fin and the whitish saddle patch located just behind it. At the time, natural marking studies had been done on large African mammals—rhinos and elephants, among others—but the technique was not applied to marine mammals until Bigg began his study and, at about the same time, Roger Payne of the New York Zoological Society began studying southern right whales. The orca study has become the most complete photographic record of any marine mammal in the world. Using that method, 190 whales have been identified in a northern community, centered near the north end of Vancouver Island, 88 more in a southern community that inhabits the area around San Juan Island and southern Vancouver Island, and 80 have been identified as a transient community that roams some nine hundred miles up and down the coast. In each community, the family groups called pods, centered on the mother, have been identified and given alphabetic designations, with a numbered suffix for each individual whale.

When Ken started, he knew little about orcas, other than their taxonomy. The order cetacea (whales, dolphins, and porpoises) is made up of two groups of living whales, baleen whales, those with large plates in their mouths, used like a colander for scooping up organisms, and toothed whales, of which there are some sixty-seven species, including dolphins.

The orca is classified as the largest oceanic dolphin. Ken was ready to believe the stories about their ferocity toward humans, which may have been extrapolated from their feeding habits. In the navy, where his interest in whales had developed, the diving manual had described the orca as a "ruthless and ferocious beast." If an orca was spotted, the diver was advised to get out of the water immediately. Such descriptions were probably based on observations of them hunting in a pack, like a wolf on land, and feeding on other whales. In 1874, for example, whaler and naturalist Charles M. Scammon described an attack by killer whales upon other whales as being like "a pack of hounds holding the stricken deer at bay," clustering about the animal's head, seizing it by the lips, and hauling the bleeding monster underwater, "and when captured . . . they eat out its tongue." A 1954 *Time* magazine article referred to them as "savage sea cannibals . . . with teeth like bayonets." The article went on to describe how orcas in Iceland had destroyed thousands of dollars worth of fishing tackle and the government asked the United States for help. Seventy-nine bored soldiers, stationed at a lonely NATO air base on the subarctic island, responded. "Armed with rifles and machine guns, one posse of Americans climbed into four small boats and in one morning wiped out a pack of 100 killers," stated *Time*. The article did not then question, which animal is the more ferocious?

Orcas do hunt in packs. They have often been called the wolf of the sea, and the accounts of their methodical pack hunting have been well-documented. What has been found to be true of the wolf is also true of the orca, which is that they eat because they are hungry and rarely eat more than they need to survive. The orca is an opportunistic hunter and tends to be a specialist. The community in and around Puget Sound eats fish, particularly salmon, and seems to follow the seasonal salmon runs, as do those in northern Vancouver Island. The transients seem to prefer sea lions and seals. In open sections of the Antarctic, the orcas eat minke whales, while near the ice they usually feed on seals. In the Peninsula Valdes in the South Atlantic, the orcas prowl the sea lion rookeries along the beaches, feeding on sea lion pups. They have also been seen hunting humpback whales, which are about twice the length and four or five times the weight of a mature orca; gray whales; and even the blue whale, the largest mammal on the planet. However, despite their skilled hunting, there are no accounts of an orca actually attacking and killing a human and never any reason to fear them as a man-eater.

Ken's study began on April Fool's Day 1976. He spotted his first pod six days later. As he poked his boat alongside them, the orcas showed a mild curiosity, but for the most part they went on about their business. He did the same. He snapped photographs. He also stopped thinking of them as killers. Looking back on that moment today, he considers it remarkable that they allowed any boat to get near them, since the whales he followed that day had already been "exploited" or "cropped," the favored terms for capture. From 1962 to 1973, according to his own review of capture data, at least fifty-eight orcas were removed from the waters around Puget Sound.

One month before Ken began his study, Don Goldsberry, working for Sea World and using boats, aircraft, and explosives, herded six whales into Budd Inlet in Puget Sound near Olympia. At the same time, the state legislature was debating the possibility of a sanctuary for whales in Puget Sound, so driving the whales into the inlet was perhaps not the most brilliant public relations strategy. The state government filed suit to stop the capture, three whales escaped, and a resolution was passed calling for a halt to killer whale captures. The remaining whales were ordered released and Puget Sound became an unofficial "sanctuary." Since that time, Iceland has been the site of most orca captures for the world's oceanariums and aquariums; that's where Bjossa and Finna were caught.

For the next several years, Ken funded his study largely out of his own pocket, using the salary from winters spent on the *Regina Maris,* a square-rigged ship out of Gloucester, Massachusetts, on which he served as chief scientist for nearly twelve years. In 1979, he started a Whale Museum in Friday Harbor. His first exhibit was the headless skeleton of an orca that had washed up on the beach of the Olympic Peninsula. He hoped to use the proceeds from the museum to fund research, but it soon grew popular enough to require a board of directors. Unfortunately, as one biologist who still works for the museum told me, "Ken is one of the best field biologists anywhere, but he isn't so good with a board of directors." Among other things, the board was uncomfortable with the car Ken was driving, a black, 1962 Chevrolet Impala, nicknamed the "Low Rider." It had more than one hundred thousand miles on it when purchased and made a disturbing, low-pitched grumbling sound when he drove through Friday Harbor. So, Ken was fired. Or, he quit. Or, most likely, some confusing combination of the two.

The Whale Museum expanded and prospered; nearly twenty thousand visitors pass through each summer. Ken, taught as a navy pilot that you can't blame a crash on someone else, started over on the west side of the island. Eventually, he gave his new organization a name and for the past six years has relied on volunteers from the environmental organization "Earthwatch"—six ten-person teams per summer—along with his girlfriend and two more or less permanent volunteers. On a budget of about fifty thousand dollars, he maintains two boats, food, cameras, and computers. "If you give up control of your life, which most people do, well, that's it," he said. "That's everything."

One early morning a few days after I arrived, I was standing on the porch of the volunteers' house stunned by the view. The air was cold, but the sun was brilliant, although the house was still in shadow. I could see whales blowing—"kwhoof, kwhoof, kwhoof"—minute puffs of vapor on the horizon. They were traveling north in the strait, moving fast. The water was a deep, ultramarine blue with choppy waves about a foot high. Suddenly, Astrid van Ginneken, one of the more or less permanent volunteers, appeared on the porch beside me. An assistant professor on the Faculty of Medicine at Irasmus University in Rotterdam, with a Ph.D. in medical informatics—the use of computers for medical diagnosis—she'd been spending six to eight weeks each summer at Ken's research site since 1987, at her own expense. Two front teeth protruding through closed lips gave her a mysterious, perpetual smile. She looked slightly rumpled, like she'd dressed in a rush. She said the whales had awakened her.

"That's J8 out there," she said. "I could hear her wheezing."

Wheezing? I could barely see them. She went to the spotting scope, looked through for a minute just to make sure she'd heard correctly, and went inside to record the sighting.

Within the hour, there had been another report of whales off False Bay, to the south. We used a small rubber dinghy to reach the Trimaran, which was named *High Spirits*. Before pushing off from the small pebble beach, I removed my sneakers, tied them together, and looped the laces around my neck. It was time to go whaling. When we reached the Trimaran, feeling very much the old salt, I didn't put the sneakers back on. I later asked myself, if I was smart enough to wear a white cotton cap to keep the sun off my head and out of my eyes, mightn't I have suspected that I should provide the same protection for my feet?

We tucked the dinghy along the front of the Trimaran and let loose the anchor. Ken had taken the Boston Whaler off for repair. His girlfriend, Diane Claridge, a short blonde, with owlish eyes, a deep tan, and brilliant white teeth, was at the helm. She came to the San Juans in 1988, after meeting Balcomb on the *Regina Maris* several years before. Also on board: Astrid; Dave Ellifrit, the second more or less permanent volunteer, a red-haired, but quickly balding, twenty-three-year-old student from the University of Missouri, who spends each summer with Ken; Dave's twenty-eight-year-old brother, Paul, who came along this year to cook and tend the garden; and four other Earthwatch volunteers.

We cruised past a reef surrounded by a large bed of kelp and water infested with jellyfish, before making a wide turn to the south in Haro Strait. On the far horizon, the rugged, snow-covered peaks of the Olympic Mountains poked through a thick, comforting blanket of clouds. The sun was bright and hot, but the water was cold—about forty-five degrees—and so was the air over it. Even a slow cruising speed of five miles per hour forced everyone into sweaters and jackets. Much of that day would be spent taking off or putting on clothing, as the boat either slowed or accelerated. Twenty minutes later we were passing the Lime Kiln Lighthouse on the west side of the island when Diane said quietly, almost to herself, "whales." It was the 1990s version of "Whale, Ho."

Dave responded, "Yep, I saw that too," while two of the volunteers made a note of the sighting, which neither they nor I had seen, in the log.

Diane steered the Trimaran to starboard, to get outside of the whales and give them some room, then shut the engine off. We drifted. The sun again felt hot. I stripped off my jacket, intently watching forward. Dave, looking through binoculars, said, "Here they come," and suddenly a whale surfaced several hundred yards to our left front.

"Who is it Dave?" asked Diane.

"So far, it's a black dot," he said.

And it was gone. Dave said, "It looked like K20."

Diane cranked up the engine and turned the boat perpendicular to the shore, which was about a half-mile distant. Houses dotted the brownish green hillside, among scattered fir trees and a thick splash of orange poppies, which looked as though they had been layered by a painter's brush against the side of the island. Dave raised his binoculars. Another whale surfaced several hundred yards away, which Dave identified as K1, an adult male whose dorsal fin had two notches in it, so symmetrical they looked as if they had been carved by a knife. When I mentioned it, I was told K1 had been captured in 1973, and the two notches had been made surgically, so he would be easier to identify, like banding a bird. Then, farther away, K5 surfaced right through three seagulls, who had been sunning themselves on a piece of driftwood. They scattered like splinters in the wind. K5 was also an adult male, with a dorsal fin free of notches, but which leaned to the right.

We drifted aimlessly, as if waiting for the real game to start. The goal was to maneuver the boat so that a proper identification photograph could be taken. Each summer, they shoot one profile photograph of each orca's dorsal fin, left and right sides, from approximately the same position—about one hundred to two hundred feet away—using black and white Ilford film (400 ASA) at a shutter speed of 1/2,000 of a second. They had already photographed K-pod, so, while they logged their sightings, they followed them with the hope of finding L-pod, whose photographs they needed.

One-half hour later and without a whale in sight, we were further south along the west coast, off False Bay. To our left, closer to shore, was a huge cabin cruiser named the *Blue Mist*, with five sailboats gathered around it, like a mother duck with ducklings. Diane grabbed her binoculars, focused on the front of the cabin cruiser, and said, "There's a nice, big group, real tight." We could see them blowing—"kwhoof"—and heading north. Diane pulled back the throttle, then cut it completely and grabbed her camera. Dave had two cameras around his neck, one with black and white film for official photos, the other with color for his own shots. We waited, bobbing in the water, squinting, and—"kwhoof"—two whales surfaced right in front of the boat. Dave quickly swiveled and shot twice, calling out the identification—"L61 and J18." Charlene, a volunteer in a big sunhat and black sweatshirt, scribbled the ID in the log. Just as quickly as they'd surfaced, they were gone. Then to the front, four or five whales could be seen bearing straight for the boat. Another unexpectedly surfaced under the bow, announcing himself with a huge exhalation—"kwhoof." I could feel the mist and smell it, heavy and oily, as cameras whirred on motor drive. The orca dove beneath us, while everyone scrambled to the opposite side and waited for him to resurface. "L44," Dave called out, a ten-year-old male with a short, rounded dorsal fin. We waited, but he never reappeared. Whales were everywhere around the boat, some close, some distant and too far for pictures. The water was flat and calm and dotted with dorsal fins and "kwhoofs." One whale, unidentified, did what could be described as a barrel roll. Charlene, keeper of the log, asked Diane, "What is it when they do that?" Diane paused a beat and said, "That was rolling over." Another pause. "We don't have names for everything."

She wheeled the boat around and we headed north. We were still about a half mile from shore. Suddenly, off to the left about a hundred yards, a whale breached, hurling itself completely out of the water and reentering with a heavy splash. It's the only time a whale completely leaves its environment, diving into the air the way we might dive into the water, turning and twisting, ebony black and a flash of pearl white etched against a hazy blue sky.

A few weeks before, I'd read an article in *Scientific American* entitled, "Why Whales Leap." The author, who had earned his Ph.D. in zoology by studying the behavior and ecology of humpbacks, concluded that, while his findings did not indicate any single clear function, the breach often served to "accentuate other visual or acoustic communication," as people might jump up and down to emphasize a communication. So, it was possible the whale I'd just seen leaping and spinning was in the middle of an argument. On the other hand, such an explanation reminded me of Alfred North Whitehead, the physicist turned philosopher, who once said that people could have extensive scientific knowledge of the sun, the atmosphere, and the rotation of the earth and still fail to appreciate "the radiance of the sunset."

The researcher considered, but rejected, the "somewhat blurry concept of play," but play behavior, in fact, is found in most mammals and even appears in over forty species of birds. One of its main characteristics is that it is rewarding in and of itself. It's not done for survival, nor for caring for the young, avoiding danger, procreation, or finding food, or even for communication. Thus, only recently have biologists begun to study play behavior in animals, and to imagine it might even exist. They have been forced to, because the closer they studied certain animal behavior, the more time they found devoted to play. Now, they are beginning to reason that it must be important to growth and development. In humans, more than one philosopher has considered art an adult form of play. It has also been deemed essential in stimulating children to think, invent, adapt, create, and even use language fluently. Play, in fact, is an expression of intelligence and indispensable for creativity. If biologists are now willing to admit animals play, is it such a large leap to imagine that they also play with their minds?

Perhaps. Such a leap can easily earn the scorn of biologists. Yet, as I watched the orca, it was impossible to keep myself from imagining the pure pleasure of leaping into the air and soaring with them. I even imagined that they were conscious of how they looked, like dancers, sensing if they had fully left the water or had their tail dragging. When they were free of the water and poised at the peak of their jump, I believed they felt as I did as a child when I threw myself off the high dive, out of touch with the familiar, supported by nothing, tickled by cool air and goose bumps, bracing for the impact and hoping I could make a bigger splash than my brother.

My leap of imagination was interrupted by Dave, who called out, "We want to head over toward that calf," and Diane began maneuvering for a photograph of a newborn. There were several in that particular pod. A whale surfaced behind the boat—L10—an adult male estimated to be thirty-two years old. He was making a sucking sound, like an old man gasping for breath. He dove and disappeared. We slowed. Dave was poised with his camera. We waited. Suddenly, the water to our left front began boiling with thrashing tails and pectoral slaps and, in the midst of it all, mother and calf quickly surfaced, mom pushing the tiny calf along on her forehead, and they dove, too quickly for any photographs. We followed the group, which was only about fifty feet to our front. Diane turned the wheel over to Paul, so she could be free to photograph, and then quickly called out, "L83 on frame 21,"—a photo of one of the calves, which was promptly recorded in the log. A big group surfaced to our right—five, six, seven whales—and Dave called out visual identification—"L43, L72, J1, L62, J5, and . . . I missed one." "You *missed* one?" snorted his brother, Paul. Dave looked embarrassed. Then, suddenly, the whales were gone. The water calmed. It was eerie to have them drop out of sight so quickly, but we meet them at the surface of the water. They can disappear from our world when they choose.

By midafternoon, we were in the same general vicinity, having traveled in several large circles in a northern direction. If the whales were as curious about us as we were about them, I'm sure they would have been dizzy by then. From a distance, we started following a group of whales headed north. The whales were probably following the salmon schools. In captivity, to maintain weight they eat about 3 percent of their body weight each day. In the wild, they might eat slightly more, which, for an average-sized male, could be three hundred to four hundred pounds of fish each day. While we followed, we passed the lighthouse, then the Whale Watch Park, established by the state of Washington in 1985. About twenty-five people were standing on shore, looking like small potted plants.

Diane suddenly announced, "The goal is to find L4 and the calf." The hunt was on. L4, a female believed to be about fifty-four years old, had been seen with the calf four times, and it had been assumed she was a grandmother, taking care of someone else's baby. Now, they were certain it was her calf and they needed family pictures. We picked up speed and continued north. The air was cold. Jackets were pulled on. Ahead of us were two lemon yellow kayaks, tiny in the water. Suddenly—what the hell?—a whale surfaced just behind the boat, rising out of the water like Captain Nemo's ship from *20,000 Leagues Under the Sea,* and dove beneath us. Dave did a jig on the deck and shouted, "What a whale." It was J3. Just as suddenly, a line of about a dozen dorsal fins appeared in front of the boat, rising and falling regally, also headed north. They passed the kayaks. The kayakers stiffened, then one of them began paddling madly after them. For a few seconds, the calf could be seen just in front of him as he pumped, leaning his body into each stroke, but after thirty seconds he gave up, exhausted, and rested his paddle across the bow. In just thirty seconds, the whales, who can swim up to thirty miles per hour, were already distant.

Long cirrus clouds stretched across the sky from north to southwest, diluting the late afternoon sun. We could see whales to our front and some closer to shore. To the left, toward the open strait, the water was deep ultramarine, but between us and shore, picking up reflections from the rocky sandstone, it was a hundred shades of brown and green. Beyond the Whale Watch Park, the gray and brown sandstone cliffs rose steeply from the water, then flattened out and gave way to a small beach. People on shore were squatting among the sandstone rocks, some almost in the water, and children screamed a high pitched, "There they are," as a whale smacked the water several times with his tail and another surfaced, draped in kelp. Another whale breached, with a thunderous crash of water that echoed off the cliffs. The performance was rewarded by a burst of applause from shore.

We continued north, watching the whales from a distance. An hour passed and during much of that time a whale followed behind the boat. I heard him blowing as he cruised on the surface, sometimes sounding like the intake on a whistle, other times as if he were gasping for breath. We crossed a channel of rough, green water, choppy from the tide, until we were off the point of a fifty-foot sandstone cliff called Kellet Bluff, approaching a large group of whales to our right. Dave scanned them with his binoculars. Identification began. It was a mixture of whales from J and L pods, twenty-three in all, boiling the water with pectoral slaps, tail lobs, barrel rolls, and an occasional breach. A mob scene. Rainbows formed in the mist. Imagine a crowd at Forty-second Street and Fifth Avenue in New York City suddenly hugging and touching and acting as if they were happy to see one another. Perhaps twenty yards away, two whales wrestled in the water. Just as Dave identified one of them—"J6"—two volunteers shouted, "Sea Snake," and a long, pink, shiny penis skimmed the water's surface. Everyone smiled.

Astrid was beyond smiles. She looked beatific. "That's J6 with L21 again," she said. "He was after her last year. Three summers in a row. Isn't that something?" She spoke as if she were talking about her own adolescent children coping with racing hormones. She raised her eyebrows slightly and looked at me carefully. Whenever Astrid explained something, she would look the person carefully in the eye. If she determined you shared her enthusiasm, she'd giggle, satisfied, but if you appeared skeptical, she'd disguise her enthusiasm. The shout went up again—"Sea Snake." We turned to see the pink organ reflecting the late afternoon sun. "J6," Astrid said gently, practically putting it to music. She lowered her voice to a confidential tone. "It's going to be a great child they have. If L21 has a calf, we know the Dad."

The event was entered into the log as, "J6 hot for L21."

We were still short of the goal. A twenty-foot sailboat, a few hundred yards to our right, was also following the whales, and fifteen or twenty minutes passed as we all drifted slowly north from the bluff. The whales started to spread out. Suddenly, a five-foot swell rolled through from the northwest, the wake of an enormous car ship that had passed through the strait. We easily rode up and over it. To our right front, perhaps a hundred yards, four whales—J18, L10, J8, and the frisky J6—dorsal fins sticking straight up, rode the swell for five to ten seconds, surfing, as if they were racing one another, then dove beneath the surface. Everyone on the boat was quiet. "Meditation and water are wedded forever," wrote Melville in *Moby Dick*, and perhaps meditation is prompted by tiny events like the one we had just seen; waiting, straining and, without warning, a sudden glimpse beneath the surface of things.

"Wow," said one of the men in the sailboat. The word remained suspended in the quiet between the two boats.

"Just . . . wow."

Diane peered at me and said, "That is the *first* time I have ever seen *anything* like *that*." It was the briefest of moments, but it seemed to speak of all the mystery in the world.

GONE WHALING

The sun was low. The air was cold. It was clear enough to see the snow-covered summits of ten-thousand-foot Mount Baker to the east and fourteen-thousand-foot Mount Rainier, in the Cascade Mountain range fifty miles away, to the south. On the horizon across the strait, the Olympics were purple, white, and jagged, no longer softened by clouds. Kellet Bluff was behind us. Ahead, five whales were cruising. Dave quickly identified two of them as L4 and L86. The goal was in sight. Within fifteen minutes, Diane had maneuvered the boat slightly ahead and to the left of the whales, with the sun behind us, the perfect picture taking position. Cameras were pulled from cases and focused. The whales dove and surfaced in a regular rhythm a hundred feet away: Mom— L4—formerly thought to be grandmom—and her baby— L86—who looked tiny, but at birth was probably more than six feet in length and weighed four hundred pounds. What sound did mom use to call her calf? Surely it must be more poetic than L86? They dove, three, four, five times. The calf had an orange tinge, common in newborns, presumably because its blood vessels are much closer to the surface, and was so close to the mother it must have been touching or feeling her or at the very least being carried along in her wake, perhaps matching its breath to hers. Watching them, I remembered how, as a child, I would creep through the dark to my parents room, climb between them in their bed, and rest my hands first on my father's back, then my mother's, feeling them breathe and watching my hands slowly rise and fall with each breath. I would ride in their wake, comforted, through the night.

Five people, knees bent to absorb the swells, were lined along the starboard side, paparazzi ready to catch a glimpse of the latest big thing. Five cameras were at the ready. Mom and

her calf rose slowly, the cameras began firing. "Phwiip . . .
phwiip . . . phwiip . . . phwiip . . . phwiip."

In 1976 Ken and a few other re-
searchers took fourteen thousand black-and-white photographs of these whales. After that, he stopped counting.
From that enormous catalog of photographs, family histories
have been pieced together and inferences have been drawn.
Typically, female orcas give birth to their first calf at about
thirteen to fifteen years and have about five babies during the
next twenty-five years, although mothers eleven years old
and more than fifty years old have recently been identified, so
the estimates are always subject to revision. A joking expression I heard frequently among orca researchers was, "recall
my thesis," which was used just about every time a new bit of
information was exchanged. Several females have had calves
at three-year intervals, which might be about as fast as they
can have them, since the gestation period lasts seventeen
months and the calf nurses for about a year. The offspring,
even after weaning, continue to have strong bonds with the
mothers; grandmothers are in almost continuous association
with daughters and granddaughters, and males apparently do
not disperse, but remain by the mother's side throughout her
life. She is also the one who lives longest. It is estimated that
females live as long as seventy years, males fifty years. The
mother is the glue that holds the pod together.

In late summer and early autumn, the resident pods of the
southern community come together in what Ken calls a
"greeting ceremony." Much like the mob of activity seen that
day off Kellet Bluff, there is a lot of close contact and sexual
activity. Southern resident males have been seen sexually ac-

tive with females, other males, and juveniles within their pod and their community, but they have never been seen to mate with the "transient" pods or with whales from the northern community. But, just when Ken thinks he knows them, and is perhaps a bit complacent about it all, something happens. In August 1990, Prentice Bloedel was taking an ocean survey twenty-five miles west of Vancouver Island, and he ran into a group of perhaps forty whales. He thought it was a group from L-pod and photographed thirty-two of them. Later, it was found that the photographs showed thirty-two unidentified whales, traveling in a large pod like "residents," eating fish and similar in physical characteristics, with smoothly rounded dorsal fins and a distinctly shaped saddle patch. Shortly before that, another researcher, Jim Darling, who now works at the Pacific Biological Station in Nanaimo, was out looking for humpback whales west of Vancouver Island and came upon about twenty or thirty orcas. He, too, thought it was L-pod. He made a twenty-minute audio tape and sent it to John Ford at the Vancouver Aquarium, who immediately said to himself, "Recall my thesis, because L-pod has changed its dialect." Darling had also photographed three of the whales. As it turned out they, too, were previously unidentified, and none matched with the photos taken of the other new group. Finally, several people sent photographs and an audio tape from the Queen Charlotte Islands, which hinted at the possibility of yet a third group. In total, there could be a hundred or more previously unidentified orcas—the "mystery whales."

The first two groups were well within the known offshore range of "resident" whales from both northern and southern communities. Ken suspects, but has not observed, that when the northern or southern inshore residents head for deeper water, they may occasionally meet with those offshore whales, intermingling with them in a "greeting ceremony" and, perhaps, interbreeding. It would explain the similarity between the northern and southern resident whales. The northern and southern communities might not be separate communities, as such, but smaller parts of a much larger group. Offspring would stay with their mothers and her community. To an outsider—like Ken—the pod structure would appear to stay the same. The difference could be that, by mixing with offshore resident whales, the gene pool is expanded.

Ken's observation also reminds us that it's a big sea and that we know a lot less about it than we might think we do. We pick up bits and pieces about orcas and, as with all our reasoning about such animals—about each other, about the world in general—we extrapolate from those limited pieces. Most of the time I spent with Ken's group following whales on the strait, we saw nothing. They were going on about their lives, plowing their way through changes in temperature and light beneath the surface, perhaps noticing a change in salinity or a change in pressure as they dove deep and then rose to the surface to take a breath. We grasp that brief moment, construct a system, and say, This is what we know, when that system may be no closer to the truth than the sandstone backdrop at the aquarium.

"Why do orcas surf?" I asked Balcomb. We were in the living room of the waterfront house the following morning.

"Beats the hell out of me," he said.

Four large picture windows faced the strait, which stretched crystal blue in all directions. Diane was on the deck, looking through the Nikon fieldscope. Those who had been on the boat the previous day were expected to stay in and do the office work, identifying and cataloging photographs, among other things. I joined them. In doing so, I also avoided walking. Overnight, my feet had swollen from sunburn, the natural result of having removed my Nike "Force" High Top sneakers on the boat and exposed pinkish feet to an intense sun for about twelve hours. They looked like casaba melons, which are plump, with a yellow rind. Normally, my feet are bony, but the bones and my ankles had completely disappeared beneath the swelling. For decorative effect, three horrible yellow blisters had erupted on my right foot, each as big as a quarter. Everyone agreed they'd never seen anything quite like it. To care for them, I had to soak my feet in vinegar, cover them with tea bags, and dip them in the forty-five-degree water of Haro Strait, which made them completely numb after five minutes. At that moment, I was sitting on the couch applying Neosporin.

Astrid came into the room.

"Did you have a good encounter?" she asked me.

It was a frequently asked question, not just of me, but of all volunteers. I'd been asked the same thing a few days before by another volunteer and I'd simply answered, "Sure," although I wasn't quite certain what it meant. I'd heard another volunteer being asked, "Have you had an encounter yet?" Also, I'd heard Astrid tell one of them, upon our return the day before, "We had a great encounter."

"What do you mean by encounter, Astrid?"

She narrowed her pale blue eyes.

"You know," she said, smiling.

As best as I could determine, an encounter was defined as seeing the whales close enough to feel their breath or look them in the eye or have your feet tingle. I think what Astrid meant was, a moment of empathetic illumination, although she joked about it, rather than let anyone, particularly a relative stranger, suspect her affection for the whales went beyond the bounds of acceptable scientific curiosity. I thought she was very serious, since such an encounter would mean you had crossed the line from observer to one who understands or, perhaps, had seen the reality behind the framework. If that was the definition of encounter, which I'd just made up, then I hadn't had one. I hadn't felt the illumination. I hadn't crossed the line.

"Well, I saw them surf," I said.

"That's good," she replied, still smiling. "That's close."

Ken came in the room with a cup of coffee and sat on the couch. Astrid turned to him.

"I think you have to start regulating the number of whale watch boats," she said.

The day before, they had seen boats from Victoria, Bellingham, and Friday Harbor.

"When we started, a biologist friend of mine said he thought humans around whales were just like flies around a horse," said Ken. "Minor irritants at best." Now, humans around whales had become a recreational industry, which was disturbing to some people, although there was no evidence that the whales were affected by it in any way at all.

"I think a lot of them are being harassed," said Astrid.

"Sure, *you* might think that, but what's ironic about it is this," said Ken. "When we started our study, it was right after a series of captures, so I'm sure the population was traumatized by removals and were very circumspect around boats. But, they still had to be around boats. So, the whales would identify which boats were threats, which weren't. Today, certainly, they are not as circumspect. They are more vivacious, swimming up to the boats, poking into bays. They never would have done that in the past, because that's where they were captured. Really, you look out here in the strait and you don't see any stress problems. Or, I should say, I don't see any stress problems and that's based on my own historical observations. If I were arriving here today, I might look out here and say that having that many boats around is bad. But, the whales out here are doing very well. If they're being harassed, it must do them some good, if we can surmise anything from the fact that they are thriving."

From the porch, Diane said loudly, "Whales," which was like a fire alarm bell for Dave, who came from the kitchen. He scanned the water.

"No, that's a wave," he said.

Diane looked through the scope again.

"No, it's whales, lots of them. Heading south."

She was right. The volunteers on the boat shift started to pack their gear. I applied more Neosporin to my feet. The room in which we were sitting had been set up to steep the volunteers in whales. On the coffee table was Ken's own book, *The World's Whales: The Complete Illustrated Guide*, coauthored with Stanley M. Minasian of the Marine Mammal Fund, and a copy of *Marine Birds and Mammals of Puget Sound* by Ken and Tony Angell. Five dorsal-fin silhouettes had been placed in one of the windows, to help the volunteers practice identification, like coast watchers in the Second World War trying to memorize silhouettes of Stukas and Messerschmitts. A dramatic photograph of a breaching humpback whale hung on the wall. On the piano sat a grinning, 160-pound orca skull.

Ken explained that when he was in the navy he was stationed near Taiji, Japan, the center of Japanese whaling. He was in a taxi, passing by the docks, and saw the carcass of an orca, which had just been towed in by some fishermen. He yelled "stop" at the taxi driver, who didn't understand English, so Ken jumped out while the cab was still moving. A researcher from the local aquarium, who spoke English, happened to be there, and Ken asked him to explain to the fishermen that he'd like to have the skull of the orca. The fellow translated. No, no, no, no, the fishermen said, shaking their heads. They intended to sell the teeth for tuna lures. Twenty dollars each. On the spot, Ken bought the teeth for five hundred dollars. With the teeth, came the skull. He placed the head, which was covered with muscle and fat and weighed a bloody 300 pounds, in the trunk of the cab, took it to the aquarium, and had it lowered into a tidal pool. One year later, the skull had been cleaned and polished by the organisms in the pool and weighed about 160 pounds. Ken shipped it home as part of his research collection, labeled household goods.

I hobbled to the basement. Dave was going to show me how to photo ID. Downstairs, two rooms faced the strait, although the view was partially blocked by overgrown blackberry bushes. One of them was Ken's office, where he had an old oak schoolteacher's desk, a matching oak file cabinet, and a five-foot oak bookcase stuffed to overflowing with, among other things, the *Handbook of Marine Mammals*, volumes one through three, *Whales, Dolphins and Porpoises* by Kenneth Norris, and dozens of copies of the *Journal of Mammology* and *Oceanus*, the international magazine of Marine Science and Policy, published at the Woods Hole Oceanographic Institution. The rustic feel was offset by the nearby fax machine, NEC Multisync computer, and a Hewlett Packard Paint Jet XL Printer. An oil painting done by Ken's grandmother hung over his desk. It appeared to be the view from his house looking south along the shore of San Juan Island, but painted years before Ken had moved there. A photograph of Michael Bigg, the British Columbia marine biologist, was tucked in the frame.

I found Dave crouched over a light table in the back room. He's thin and has a faint beard. Charts of orca dorsal fins were on the wall above him and ring-binders full of photographs were on the shelf. I sat on the stool next to him.

"The best way of getting an idea of how this is done is to do it," he said, handing me a roll of film. So, I started. Using the negatives, I looked first for the bulls, the easiest to identify with their large dorsal fins, up to six feet in height. Females have smaller, smoother dorsal fins and it's much more difficult to tell them apart. The calf pictures, which had been difficult to get, were also particularly difficult to identify, because the saddle patch was faint. I tried comparing the negative image with the chart of photographs. Was that dorsal fin the same shape? Was that a nick near the top? While I struggled, Dave described some life histories. There had been six calves the year before, of which four survived, including a set of twins. About half of the calves born in the wild die in the first year. In J-pod, the last calf had been born in 1988, so it was possible something was wrong with them. However, K-pod had done the same thing, gone seven years without a calf, then started pumping them out—seven calves, of which five had made it, and only two of those from whales that had previously had babies. K14 had a calf in 1988 when she was eleven years old and another last September, but neither had survived. Unfortunately, they don't know why they didn't survive.

"We don't know a whole lot about their mating habits," Dave said, as he peered at his strip of negatives. "The theory popular now for some whales is sperm competition—loosely translated as the 'big ball theory'—which means that several big males will mate with the same female and the most sperm wins." They inflict some nicks and cuts during mating, but compared to how powerful they are, it's relatively gentle.

I continued to peer at the strip of negatives and asked him how he had developed an interest in whales. He hesitated, blushed, then confessed that he'd seen a movie when he was a kid based on the life of Namu, an orca at the Seattle Aquarium who later starred in a National Geographic film. In 1984 he'd taken a trip to the north end of Vancouver Island to see Namu's mother, who after the capture had followed Namu all the way to Campbell River, about halfway down Vancouver Island. Back home in Columbia, Missouri, Dave had studied whale photographs for years, before coming to work with Ken. When he did, "they knew I knew my whales." After a week, they stopped checking his film.

He was concerned about my film, though. He checked a few, didn't say anything for a minute, then said, "It wasn't until last summer that I could ID females from the boat. I had to look at a lot of negatives before I could do that." He was trying to reassure me.

"Sometimes you have to ask, why am I doing this?" he said. Although I hadn't asked, I immediately confessed I had been thinking about it. "I mean, I'm doing it because I like the animals, they're my favorite thing in the world, but we are talking about eighty-eight animals and I'm not sure that knowing them does anything," he said. "People are starving, and we're taking pictures of whales. But, right now, I hope I can keep coming back every summer."

When I'd finished, he began to check the work. The first two frames were half right. There were two whales in each frame and I'd identified one of them correctly. He scanned a few more frames, murmuring, "That's right, that's right." Those were the bulls, with big dorsal fins. Then came a long string of "that's wrong" and "that's wrong,"and he pointed out nicks I hadn't seen, subtle variations in the saddle patch I hadn't noticed, a delicate finger of white just slightly different from another, a thin ridge of black in the saddle patch, or a tiny indentation on the front of the dorsal fin. In one frame, I had identified the celebrity calf—L86. I must have had a hangover from all the excitement on board the *High Spirits*. L86 wasn't even on the ID board. He studied another one for a long time and motioned for me to look. I examined the markings closely, compared them with the board, looked again at the markings.

"I still don't see it," I said.

"Yeah. You might not expect this one. This is kind of hard." It was L28 with L87, another calf. Dave started to look embarrassed again. "In fact, I can't see it either. I only know this because I remember I shouted it out at the time."

Just a few steps away, in Ken's office, I found Astrid entering data on the computer. She told me that on July 10, 1988—she announced the dates of significant events with such precision I was surprised she was unable to include the time of day—she suddenly knew the orcas. She looked off the side of the boat and started naming them.

At that time, she was working toward her Ph.D. She had

received her M.D. in 1985, and for her Ph.D., awarded in 1989, she spent almost four years building and evaluating a video disk computer diagnostic system for ovarian cancer. But even while working on her Ph.D., she maintained an interest in animals. One day—in the midst of a bridge game with friends—she discovered a book on whales by Jacques Cousteau. She borrowed it, finished it in a night, and started to look for more. In November 1986 she was in Washington, D.C., for a conference on medical informatics, and on her day off she went to the Smithsonian and bought three books on whales, two of them about orcas. She began to read one of them and on about the fourth page said to herself, "This will change my life."

After the conference, she went to New York City to see the beluga whales at the aquarium in Coney Island. She pestered the staff until she was allowed to talk to the director, who mistakenly thought she actually knew something about whales and gave her an entire afternoon of his time, as if he were talking to a colleague. He told her about orcas in captivity in Holland. When Astrid returned home, she called the aquarium and asked for permission to see the orcas. That was how she met "Gudrun." She sat at the side of the pool and sang Mozart arias. Gudrun squeaked back. She visited several times over the winter. Scuba diving lessons followed, along with computer searches for literature on whales and letters to researchers volunteering her services. No one responded. When she telephoned, she was passed from one researcher to another. Nobody had room. Nobody had time. When she nervously called Ken, whom she described as the "famous Ken Balcomb," on June 22, 1987, he told her he was full for the summer. She explained that she wanted to find a video disk application for orca research. He said he'd make room. She was there five weeks that summer and has returned every summer since. In May 1990 she presented a paper on video disk–based data collection for orca research and in November of that year, at the annual conference on medical informatics held in Washington, she delivered a twenty-minute paper on the formalization of knowledge in pathology. The last of her sixteen slides was a breaching orca.

In the meantime, in November 1987 Gudrun was transferred from Holland to Sea World in Orlando, Florida, as part of a breeding program. Never one to forget a friend, each year, after her stint in the San Juans, Astrid spends a week visiting Gudrun, who, on July 11, 1989, had a calf named Taima. Astrid said that the last time she visited Gudrun, she was sitting with her banana and yogurt when a shy man sidled up to her and asked quietly, "Are you the woman from Holland?" When she admitted it, he was ecstatic. "I'm so happy to meet you," he shouted. That's how she discovered she was considered the "famous lady from Holland who works with the famous Ken Balcomb and the wild whales."

"Do you have any doubts about doing this?" I asked. "It seems a bit less important than ovarian cancer."

Her eyes narrowed. "In Rotterdam, I live in a gray, twenty-two-story apartment building and the windows don't even open," she said. "Here, I have the sea and the sky and the whales. These two months are the highlight of my year."

She paused for a moment, made a few more data entries, and added, "My mother doesn't understand it at all. She's afraid of all animals. Even cats."

Later that evening, the crowd of volunteers was sitting on the porch facing the strait, eating supper. The sun was setting behind the Olympic Mountains.

It was a Wednesday, their last night. On Monday, they would be replaced by a new group of volunteers. The water was choppy with the incoming tide. The surf washed the rocks below with a regular beat. The cold, gentle breeze smelled of the reef, decaying and fertile. Vancouver Island was a deep green and black, stretching away to the north.

"See the breach?" Ken asked.

Everyone looked up. A group of orcas was about half a mile out, meandering back and forth in long arcs.

"This is the way it's supposed to be," Ken said. "The whales come by to see you off."

The whales wandered around the strait. Astrid took charge of the spotting scope. "J1 is with J5," she said. After about ten minutes, they passed a cormorant sitting on a sunken log just inside the reef. The cormorant didn't rustle a feather and was either ignoring them or petrified. At about 6:00 A.M., I had seen a cormorant sitting on that same log. When I looked again at noon, it was sitting there still. Then, one of the orcas spyhopped, sticking his head way up out of the water, and remained suspended there, perhaps looking at the cormorant, or the house. Several others started tail lobbing—"bam, bam, bam, bam"—like children playing in the bathtub.

Astrid was frozen to the scope, until she glanced up and saw how close the whales were, at which time she dropped everything and sprinted down the dirt trail to the pebble beach, where a two-person, lemon yellow kayak was stored. Charlene, a fourth-grade schoolteacher from Linden, Michigan, logkeeper of a few days before, who had never been within hailing distance of a kayak, called out to Astrid, "You promised to take me in the kayak," and ran after her. Three minutes later they appeared out from under the overhanging rocks. Astrid was in front, stroking professionally. Charlene was in the rear, thrashing a bit, trying to keep pace, but really just along for the ride. They fought the current. Four whales surfaced nearby several times, keeping their own pace. On the porch, Dave called out, "J17, J12, J4, and J13." Unexpectedly, J17 surfaced perhaps three feet from the kayak and, as he did, Astrid rested her paddle across the bow. Charlene copied her movement. The whale dove. Astrid, so careful with her emotions, daughter of a woman who is afraid of cats, held her head back and let out a whoop.

Later, after they'd returned to the porch, I asked Astrid, "Was it a good encounter?"

She smiled. Her discolored, slightly buck teeth were poking through her lips. "They were *big*," she said.

I was sitting on a rock, watching my sunburned feet heal, looking out at Haro Strait through the fir trees, where three crows sat, clucking to one another. The sun was warm, the breeze cold with the incoming tide. I was trying to sketch an orca, but what was its essence? Clouds began to cover the sun. It quickly grew colder. The crows lifted off from the fir trees, flapping and squawking, to chase a red-tailed hawk. The hawk seemed to ignore them. Its wings barely moved, ruffled only slightly by the wind. It rose higher and higher, riding the updrafts toward the clouds, always just out of reach of the crows, who seemed determined to make as much of a ruckus as they possibly could. If they couldn't force the hawk to leave the neighborhood, they'd be as obnoxious as possible, so he'd want to leave. He did, but on his own terms, unperturbed. Finally, the crows—jerking black dots by that time—fell away from the drifting black dot of the hawk. All was quiet, except for the determined wash of the incoming tide.

There is a contemporary American philosopher, David Lewis, who is an exponent of possible worlds. He believes that what makes this world the actual one is simply that it is the world we are in. In his view, there are many other worlds with other people in them, which are actual for those people. Lewis's position is known as "modal realism." I felt as if I had just seen an example of it. There, in front of me, ants were crawling through the brown grass and over my feet, although I couldn't even feel them, not the slightest tickle. The ants were in their own world. They had their own rhythm of life, as did the hawk, the crows, the orcas. We were all in the same place, but for each one of us that place was different.

I once read that human knowledge has been doubling itself exponentially every seven years since the late 1960s. That calculation was based on the number of scientific papers published in the world. Research on whales and other marine mammals has not been an exception to the flood of paper. As with any area of science—quantum physics or wildlife biology—there are those who pioneer and those who come afterward looking at more specific problems and slicing off pieces for themselves. Ken and Michael Bigg were considered pioneers, of a sort. Then, using photo identifications as a base, studies followed on "pod energetics," "focal animals," comparative feeding habits of the residents and transients, habitat use, and acoustics. A few years ago, A. Rus Hoelzel, a Ph.D. candidate at Cambridge University, wanted to study the genetics of the orcas in the Puget Sound in order to address two questions. First, what is the paternity of the animals? Despite years of observation, almost nothing was known about their breeding habits. Second, since orcas are at the top of the marine food chain, have they been contaminated with PCBs or other industrial wastes? Some marine life in the region, harbor seals, for example, had been shown to contain such contamination. Although they are not eaten by residents, who feed exclusively on salmon and other fish, harbor seals are the main source of food around Victoria for the transient orcas. Hoelzel envisioned firing, by bow, a barbed and tethered dart at the orcas, which would penetrate the skin and acquire a one-gram chunk of skin and body fat, an amount sufficient for DNA analysis. He had previously used the technique for a study of minke whales.

Hoelzel obtained a permit from the U.S. National Marine Fisheries Service to biopsy forty-five orcas in the waters off San Juan Island over a period of five years. However, the environmental organization Greenpeace, aided by the Sierra Club Legal Defense Fund, filed a suit to stop the project, claiming an environmental impact statement must be filed before the research could proceed. Then, the Whale Museum in Friday Harbor, whose research director had supported the project, issued a statement opposing it, after membership and financial support dropped when news of the research reached the public. Of course, the Whale Museum also has an "adoption" program, in which people purchase an adoption certificate and photograph of an individual whale. Astrid adopted one—L82—and named it Kasatka, which is Russian for killer whale and *darling*. In light of that, the only mystery was why they seemed surprised when the same people who had adopted "Ruffles" and "Ralph" did not like the idea of Ruffles and Ralph being the target of darts. Ken, who had nothing to do with the research, found himself jeered in his boat off San Juan Island by people who suspected him of using a bow instead of a camera. Hoelzel eventually had to take his research elsewhere.

Apparently, the effect of all the knowledge gathered about orcas had caused people in the San Juans to ask, Is this really necessary? While some people were upset by the protests, the traditional point of such knowledge is to prompt just such questioning, to force people to wonder at the connections of things and, therefore, to live more wisely. Yet, many biologists felt it was not a protest born of wisdom, but rather anthropomorphism. Anthropomorphism is defined as "ascribing human or other attributes to a being or thing not human." It is unscientific, which is supposed to be condemnation enough. It seemed an almost obligatory question, so I asked Ken, "How do you study whales and keep from becoming anthropomorphic?"

"I'm not sure that you do," he said. "I've often felt that our point of view of animals has a misguided sense of superiority. Of course, lots of people out here talk to them and find something spiritual about them. I'm not entirely convinced that there's something spiritual involved."

We were on the Boston whaler, bobbing in the current, just about to leave for a late afternoon cruise. The air was cold, the sun bright and sharply reflected off the water. His blue windbreaker snapped in the stiff breeze. I was wearing shoes. I had also brought my own camera, an Olympus with a 200mm lens. Ken had a Nikon equipped with 300mm lens and a motor drive. He wheeled the boat around, cruised out into the strait, then idled the boat and scanned the water with his binoculars.

At the same time, he said, people—particularly Americans—have this "whole whale thing, with killer whales and sperm whales and humpbacks and it's a soft, kind of teddy bear view of whales," which, he added, was about as close to the truth as the view once held that they were man-eaters. "Some people say we should leave them alone. But hell, we have two thousand bulk carriers coming through the strait annually, we're overfishing the salmon, polluting the sound. That's not leaving them alone. We affect them whether we study them or not."

As we sat, scanning the horizon, he said he believed that he had hit it just about perfect for his place, but what he'd really like to do—and he's not sure that he can—is expand the operation to study problems with pollution in the Puget Sound and habitat degradation throughout the orcas' range, including overfishing by fleets using gil nets, some of which are large enough to encircle Manhattan Island. He could take Earthwatch-type teams for a month, spend four or five days with the resident orcas, then take a heading out the Strait of Juan de Fuca, four hundred to five hundred miles offshore, bouncing in the foggy, open ocean, to "find out about things that are alive and things that are dying." That would be the summer. In the winter, he would head south— the eastern tropical Pacific, the Galapagos, or the Caribbean—to continue his study of humpbacks. Of course, to do all that would take about $500,000 a year, a few quantum leaps beyond what he has now, which would mean fundraising and maybe a few of what he describes as "conscience grants" from industry and polluters. Realistically, that means he would have to team up with somebody who wanted to do the fund-raising.

"You have to be careful though," he said. He started to spread his hands. "Too many times, people become divorced from what they started out to be. The reason for being becomes divorced from the fund-raising operation. Keeping the fund-raising operation going becomes the entire goal. I don't want that to happen here."

He raised his field glasses again. After a few more minutes, he spotted a pod south of the house, then quickly two, maybe three pods moving toward us. We looked around. Dorsal fins were popping up everywhere.

He increased the throttle. The wind blew his shaggy hair straight back. He began to follow, from a distance, a bull, four females, and a calf from L-pod. He looked back, identified L10—"a real bold fella"—the one who had trailed after the Trimaran, sort of gasping for breath. This time, he sounded like he was snorting.

"Do the whales still surprise you?" I shouted over the wind.

"They do," he shouted back. "I'd really like to see this place in fifty years. However developed it gets around here, with or without fifty whale watch boats prowling around, I'd like to see it. If the boats were here, that would mean the whales are still here and we're still studying them. Assuming we humans don't completely fuck everything up, which is not out of the question."

We crossed Speiden Channel. The wind was stiff and cold, the water was churned by the current, and the whales rose and fell about a hundred yards to our right. First, the six-foot fin of the male—J1—rose slowly, like the conning tower of a surfacing submarine, then his forehead broke the surface and—"kwhoof"—he exhaled. A rainbow formed in his breath. As he began to dive, the four females, with shorter dorsal fins, rose in succession until finally, blip, the calf, who, compared to the others, seemed like a cockroach scurrying for cover.

As we moved away from the protection of the island, the

air was even colder and the boat bounced in the current, dousing us with a fine spray. Half an hour passed, until we were protected again, off the west side of Stuart Island, fighting the tide. The whales were swimming close to shore. At that point, the water ripped around the corner of the island and the whales seemed to play in the tide—"working it," said Ken. They barely moved forward in the water and time and again resurfaced at practically the same spot. Clouds of rainbow vapor rose against the steep brown sandstone cliffs, now turned golden by the low sun. The warm smell of fir trees drifted from the top of the cliffs. Ken slowed the boat. We sat and bobbed in the current.

The whales suddenly went off like a fireworks display, breaching near shore, white undersides shimmering in the sun. Ken, his left hand on the throttle, right hand holding his Nikon, began to fire his motor-drive like a gunslinger, from his hip, without focusing. His knees were flexed and his weight shifted easily with the roll of the boat. "Postcard shots," he said, smiling. They were close. One rocketed out of the water not more than a hundred feet from the boat. Another did a back-flip and still another lobtailed—"bam, bam, bam." Behind us, another breach—"look"—I turned and fired my camera without focusing. Later, I found that I had frozen that burst, like catching a ballerina at the peak of a pirouette, before it was transformed into a crash of water.

The display was over as suddenly as it had started. It was quiet. One whale—J1—by now, even I could correctly identify him—surfaced close to the boat. As he passed, I thought he gave us a long look before diving.

"I think it's amazing that they are as cool around boats as they are," said Ken. "People have shot at them, they've been wrapped up in nets, taken from their families. I've seen bullet wounds in them. But more than once, I've felt that the cu-riosity is mutual, that they are as interested in us as we are in them. Maybe it's because we bug them so much."

The whales were headed north, toward the Canadian bor-der. As always, the orcas were ignoring the borders we estab-lish, although they apparently have their own, since these southern resident whales don't mix with the northern resi-dents. We followed. Ken was smiling. He looked at home.

The whales surfaced near the boat, one-by-one, each "kwhoof" blowing a cloud of vapor. A heavy oily smell swept over us. "Smell it?" Ken asked. "Whale breath. I love it."

"Why do you do this?" I asked.

We bounced along, fighting the current. We were about two miles northeast of the Turn Point lighthouse. The whales were putting distance between us and were almost out of sight. Ken was thinking. Wheeling the boat around, he said with a grin, "What else am I going to study that's still un-known?"

The whaler plowed through the current. I pulled my sweater on for the long, cold trip home. I looked back. The whales were barely visible, tiny dorsal fins headed north. Ken told me that he followed them down to Seattle one day, then followed them back out, and by the time they were nearing the Strait of Juan de Fuca, right near the docks at Whidbey Island, they started putting on a breaching show and people were standing on shore shouting and waving to them and he was taking pictures and—"phwoomp"—he went into a fog bank. He lost the whales. Unable to see, he set a compass course. Then, appearing out of the fog, the whales gathered around the boat, five or six under the bow, fifteen or twenty off to the sides, and they stayed that way, so close he some-times wished he could have pushed them away, but they stayed and after a while he simply followed them, not even looking at the compass, just following them for three or four hours through the fog, all the way home.

Another time, he stayed with them for thirty-seven hours straight, and as they were leaving Vancouver Island, headed south, J-pod formed a line on either side of the boat. Night fell. He couldn't see them, but he could hear them rising and blowing, rising and blowing, all night long. He felt just like he was part of the pod.

Johnstone

Strait

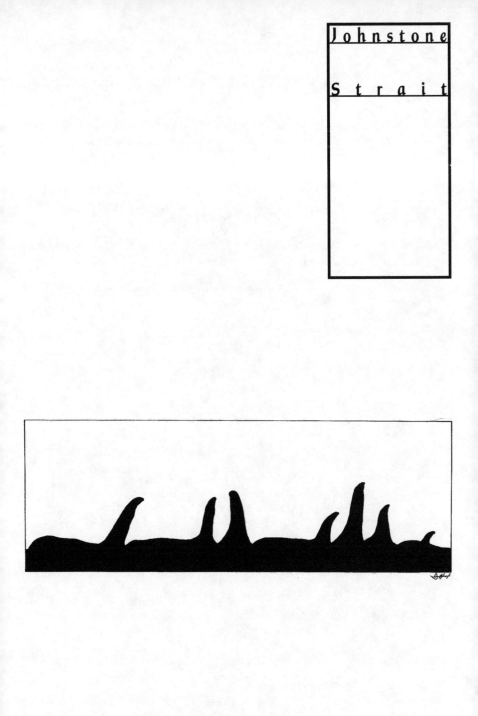

I was wet. I was cold. I was standing on the dock in Telegraph Cove waiting for Paul Spong, a man who had left the Vancouver Aquarium some twenty years ago to live on a nearby island and observe orcas in the wild. He had been described to me as the "patron saint" of the whales and a "bit of a mystic." I'd also been told he operated not on Pacific Time, but "Paul's time," so it could be a few hours or a few days before I would see him. The loosely given instructions were, hang around the dock and he'd come and find me.

GONE WHALING

Telegraph Cove, year-round population fourteen, is a boardwalk with perhaps half a dozen houses wrapped around a small cove at the north end of Vancouver Island. It had a boat landing and a campsite, where I had spent the night sleeping in some of the hardest wind and rain I had ever experienced, much of that time watching a spider on a single strand of web being blown wildly about, spinning sometimes like a pinwheel. My lightweight tent had withstood the storm about as well as the web; it had just barely escaped being blown away. Earlier that morning, as I crouched under a tarp trying to warm my hands, a middle-aged woman stopped and asked, "Is that a tent for dogs?" It is the humiliation one invites by pitching a lightweight tent near "the finest RV facility on northern Vancouver Island," where dozens of mobile homes on wheels, with names like Pace Arrow and Chieftain, had prowled through the downpour like battleships in the North Atlantic. Both the rain and fear of being crushed had kept me awake.

The rain had turned to mist, drawing delicate calligraphy on the water of the cove. In open water, twenty-five-mile-per-hour winds from the southeast had turned the water frothy with whitecaps. Small craft warnings had been posted, but despite the warnings, at the boat put-in, a man and a woman were trying to push off in a two-person, canvas "Klepper" kayak—a folding kayak that can be collapsed and carried like a backpack. He: a pink Patagonia paddling jacket with violet lining; yellow, knee-high, waterproof boots; a gray and receding hairline; and an Australian accent. He walked with the jaunty step of a mountaineer, like Sir Edmund Hillary. His bearing shouted, "Storm? What storm?" She: a blue windbreaker; blue-checked shorts; gray socks with an orange band across the top; big, round, white-framed sunglasses; and a fat, blue varicose vein on the back of each calf. She had the lumpy look of a literary agent or someone who doesn't spend much time in the sun. They each carried small, waterproof bags from their white van to the kayak, and while they packed, two other kayakers, who had been waiting for some respite from the weather, decided to call it quits for the day. They informed Sir Edmund. He responded cheerfully, "Well, we'll give it a go, and if it's too rough, we'll just come back." He strode off through the puddles to get another bag. They made nearly a dozen trips from van to kayak. On each trip, his step was more brisk, her's more and more the walk of the condemned. He asked her if she had a hat. She said, "Just this gray thing," and she put on a Sou'ester, which was about two sizes too small and sat on her head like a beanie.

After they had finished the preparations, she sat in the kayak, while he pushed them off. They paddled about twenty feet and returned. She had put her spray skirt on improperly, over rather than under her life jacket. After the oversight was corrected, he pushed off again and leaped expertly into the rear cockpit. Theoretically, the rear paddler is expected to match strokes with the person in front, but that was apparently not necessary in their case. He tried to follow her for a few strokes, gave up, and paddled powerfully forward. She poked at the water. Two bald eagles hunkered in the fir trees, chittering, as the kayak passed. Within minutes, they had disappeared around the corner of the cove into the froth of the open channel. I never saw them again.

Normally, dozens of fishermen would have been on the water, but they had been forced indoors. Earlier that day, over a breakfast of oatmeal and coffee, I'd read in the "Fishing Hole" column in the *North Island Gazette* that a 207-

pound Halibut and a 42-pound spring salmon had been caught nearby. Ina Lowe, from Telegraph Cove, was described as being "enthused" about the fishing in her neighborhood. "There's lots of fish," Ina said. But, even a monster halibut and Ina's enthusiasm couldn't convince people to brave the storm, which was why a fisherman, socked in by the weather, suddenly appeared beside me on the dock looking for company. He wore camouflage rain gear and a University of Las Vegas "Runnin' Rebels" baseball cap. I could feel him looking at me. I didn't return the stare. I didn't feel like moving. When I did, small droplets of water would drip down my back.

"You know, it used to be you really felt like you were gettin' away from it all when you came up here," he finally said. Before reading that week's "Fishing Hole," I had also read an old newspaper article about Telegraph Cove. Established in 1911 as the northern terminus of a telegraph line strung tree-to-tree along the coast of Vancouver Island, it had been described as "an archetypal little town that time forgot." A highway to the north end of Vancouver Island had been completed in 1979. Thus, I had been able to get there by taking a bus from Victoria, which dropped me in Port McNeill, a few miles north. Apparently, I was part of that which the fisherman had wanted to get away from and a reminder to him that time forgets nothing.

"When I moved west of the Mississippi River after World War II, I forgot about everything east of it," he said. "Except for pastrami. You can't get good pastrami out this way." He paused for a long time. I was staring at the whitecaps. He was staring at the side of my head. "Don't get me wrong. There are nice people in the east. Just too damn many of them."

I didn't turn toward him. I wasn't going to tell him I was

from the east, nor was I going to tell him that the folks who operated the campground had told me a huge chunk of Telegraph Cove—311 acres—had been sold to a developer. The future, apparently, would be full of people just like me, ruining the view from the dock.

"Come up here often?" I asked.

"Yeah. That's right."

"Where from?"

"Texas."

"That's a long way to come for fishing."

"Yeah. It ain't what it used to be."

"Texas, you mean?"

"*Hell*, no." He was silent for a minute, thinking. He looked away from me and out at the cove.

"Well, that too," he said.

From May through November, like the fishermen chasing the salmon runs, 190 resident orcas roam the northern Vancouver Island area near Johnstone Strait, and in turn, are chased by whale watch tour boats from Alert Bay, Sointula, Port McNeil, and others, official and unofficial. While I was waiting for Paul Spong, I signed on for a tour with Stubbs Island Charters, run by Jim and Anne Borrowman and Bill and Donna McKay out of a two-story red barn at the end of the boardwalk in Telegraph Cove. Their office also contains the smallest post office in Canada. In a hundred-day season, they take anywhere from thirty-five hundred to four thousand people, usually on the *Lukwa*—"fishing for halibut" in the local Native American dialect—a ninety-foot boat with an aluminum hull and two 540-horsepower engines. They had it designed and built two years ago expressly for whale watching at a cost of $1 million, which means that business has been good and had better continue that way for quite some time. They also own the *Gikimi*, a smaller, older wooden fishing boat, with which they started their business, first hauling lumber, then any odd job to bring in a few dollars, then whale watching. In 1982 they published their first brochure, "Discover Northern Vancouver Island," with a fuzzy idea of exposing people to the great natural beauty of the British Columbia coast. They used a photograph of an orca on the brochure's cover. It wasn't long before they realized that the whales were the only thing people wanted to see.

We were in the pilothouse of the *Lukwa*. Wearing a royal blue fleece shirt, navy blue corduroy baseball cap, and white shorts, Jim was ignoring the weather. "It's summer," he said. He was standing over an enormous, polished oak console. Five windows faced front into the fog. Rain drummed steadily on the roof. The rain squall was a green, blossoming flower on the circular radar screen as we chugged easily through the whitecaps on a gray-green sea. Jim described himself as "possessed" by the whales and keeps a thirteen-foot orca skeleton on the roof of his house on the boardwalk, but he doesn't know why other people want to go out in a boat to see them.

"What are people looking for?" I asked. "Why do they want to see the whales?" He seemed genuinely surprised by the question and said he'd never really thought about a reason. Perhaps it was a checklist type thing, where people just went around trying to see different species and putting them on a list, the way some birdwatchers do, he suggested. Often, he added, three or four boats would be out, and if one spotted a pod, the other boats soon arrived, circling, all looking at the same pod of whales. I had read of a 1977 study of tourists visiting Amboseli National Park in Kenya. At the height of the tourist season, convoys of twenty or more minibuses could be seen leaving the lodge together and double parking around a pride of lions. Nearly 80 percent of the vehicles used only 10 percent of the park's 300 square kilometers. At times, 80 percent of the vehicles in the park were viewing the same pride of lions at the same time. As with any form of tourism, it became important to see certain essential sights. One could no more visit Amboseli and not see a lion than one could visit New York City and not see Rockefeller Center. It seemed a fair question, then, to wonder, what would the personal experience of seeing a lion or an orca tell those people about the animal, beyond that they had been able to go and see it.

Jim finally shrugged his answer. "When you don't know much, you want to know more," he said. "Or see it. I think it's about as simple as that."

Most of the thirty or so passengers, wearing rain parkas and carrying cameras, were crowded into the pilothouse and the cabin below it, which made the interior feel like a hothouse. They included two Japanese, Hiromi, a husky photographer, and Mikiko, a tiny woman who had recently quit her job as a nurse at a Red Cross hospital in Kyoto because nurses in Japan are "really down"—not respected—although her parents were very upset with her decision. She had come to Vancouver Island last year, but failed to see an orca. This year, she was going to stay all summer until she did. She showed me a book by a Japanese photographer, Hiroya Minakuchi, whom she had met last year. Jim was keeping the book in the pilothouse because the cover photo had an Orca breaching and the *Lukwa* in the background. The text was in Japanese. The book title, *Orca Again,* and the chapter headings were in English. I asked Mikiko why the title was in English. "Japanese think English look better," she said.

We had crossed the strait, Cormorant Island on our left, Malcolm Island to our front. The surrounding islands were covered in soft fog and barely visible. The water was six hundred feet deep and Jim was sailing blind, but with a fair amount of confidence that he would find orcas, thanks to a small network of fishermen and other whale watchers, a sort of orca radio network. Paul Spong and his family were on that network, using call sign "Sacchi," which is Japanese for orca, and I spoke to him over the radio from the *Lukwa*. Of course, the network was not a guarantee. While standing in the small storefront on the boardwalk, I had listened to Jim's receptionist answer the telephone four times, and each time at some point in the conversation, she said, "Well, no, we can't *guarantee* that you will see whales," although Jim said that 95 percent of the time they do see them. Some people, he said, try to demand a guarantee—usually reporters—and he tries to be polite when he reminds them that the whales are not in a cage.

The rain turned into mist. We plunged ahead. The fog peeled away in wispy curtains. For an hour, Jim steadily answered questions—What do they eat? How many teeth do they have? How many are there? A middle-aged woman with a sketchbook, bush hat, and binoculars listened to him answering questions and stared out one of the pilothouse windows. It was her first time "in the wild," she told me, and she thought the orcas were "magnificent." Would we see them feeding on salmon? she asked Jim. Probably, he answered. Maybe something better. Just a few weeks ago he'd seen a group of transients go after a Dall's porpoise. "They stripped the skin off it and toyed with it for two hours before finally killing it," he said matter of factly.

"Oh," she said, holding her hand to her mouth. "Poor thing."

"Why did you feel sorry for the porpoise, but not the salmon?" I asked her.

She stared at me. Jim interrupted over the intercom, "I think we may have H-pod out here." She left for the upper deck along with the rest of the passengers.

"Why two hours?" I asked Jim, once she was gone. "Why so long?"

"Well, transients attack very carefully," he said. "I think the transients are really woosies at heart."

The cabin was empty, except for three children, who, it seemed, couldn't have cared less.

"It's great to see killer whales in the wild, but under the age of ten, after a few minutes, they'd rather be spilling crumbs on the carpet," Jim said. "My kids"—he has two, a daughter, eight; a son, eleven—"would rather go to an aquarium."

Save for the children, all thirty passengers were lined along the railing on the right side of the boat and out on the prow. The boat was a floating camera. Jim shut the engine down. We rocked in the water. The fog muffled all noise. While we sat, waiting, Jim told me that after the memorial service for Michael Bigg, who died in 1990 after twenty years of studying orcas, Jim, along with several other friends and family members, had taken his ashes to Johnstone Strait in the *Gikimi*. It had been a gray, rainy winter day, but when they rounded the corner out of Telegraph Cove, the rain diminished, the water calmed, and the clouds thinned, so that everything was burnished in soft sunlight. Orcas had been traveling north in the strait, and they slowed and swam around the boat. Jim put the hydrophone in the water. The whales were vocalizing. The sound was as clear as he'd ever heard it, since there were no other boats in the strait. As the orcas vocalized and swam around the boat, Jim and his friends cast Mike Bigg's ashes into the water.

He was silent for a moment. Then he said, "If anything was ever going to happen that would make me go cosmic, that was it. It made a believer out of me."

I joined the hushed crowd on the upper deck. The fog was thick, visibility a few hundred yards. Mikiko, in a navy blue rain jacket, was crouched alongside the rail, propping up a camera with a 300mm lens. In her tiny hands, it looked like an antitank weapon. Then, slowly, the way an animal on the road becomes visible through a fogged windshield, four whales appeared—one bull, H2, with a tall wavy fin and small notch on the front edge; two females; and a calf—about a hundred yards away off the port side, poking along through the fog. When they appeared, a man in a yellow rain slicker, standing to the rear, said, "Look," and the entire row of people along the railing turned around en masse, glaring at him, as if they were in church and the sermon was being interrupted by a heckler. He looked chastened. The people along the rail faced front again, cameras ready. The whales rose closer to the boat, perhaps two hundred feet away. The only sound was their forceful "kwhoof" and the occasional "phwiip, phwiip" of the cameras firing. Jim told me that a photographer from California once visited and shot a hundred rolls of film in three days.

Jim left the pilothouse and put the hydrophone in the water. When he turned the speaker on, very clear squawks and echolocation clicks could be heard. "This is live," Jim said. The orcas were out there, just off the port side, probing the water, looking for salmon. Could the salmon feel the pulse? Did they have a sense of when the echolocation clicks had locked on? Did they take evasive action or did they give themselves up to their fate? In front of us, one of the whales quickly dove and the water swirled with the struggle. I could sense the spasm under water, life squeezed from the salmon in a single crushing second and being absorbed by the orca. The feeding continued. For fifteen minutes the clicks rang from the speaker into the soft fog. Then, the clicks faded. They were finished and gone.

The fog was beginning to lift, but the misty rain continued. I turned my face into it. After spending the night in a powerful storm, it felt soft and playful. The nearby islands were becoming clearer, green and brown silhouettes. The boat was still.

Mikiko climbed down from her perch on the rail.

"Well, how was your first time?" I asked.

"I cannot find the words," she said.

She paused.

"I like the sea," she said. "The killer whale, they have their own language. I hear that the killer whale don't try to kill humans. If we don't try to kill them, they won't try to kill us."

"That could be true," I said.

"Have you met John Ford?" she asked, shyly, but with a buried sense of excitement, as if she were asking, have you met Elvis?

"Yes, I have."

"Really?"

"Yes. Really."

She looked at me with newfound admiration. Perhaps because of that admiration, she said she'd been told a story by Hiroya Minakuchi, the photographer of *Orca Again*. It was a true story. Would I like to hear it? Certainly, I said. Okay. It goes like this: Once there was a killer whale who was about to have a baby, but just before she was going to have the baby, she was harpooned and captured by a fisherman. During the capture the baby died. The fisherman, when he realized that he had killed the calf, let the mother go, but she had been injured and soon died. The father stayed with his mate until she died, then he returned looking for the fisherman. He followed the fisherman for a long time, through stormy seas and ice storms, through icebergs and frigid water, all the way to the Arctic. The orca finally caught up with the fisherman and, even though there were others there, he killed only that one person—the fisherman—who had killed one of them.

I was familiar with the story, of course. I felt as if I'd been allowed to listen as Richard Harris, in the role of the crusty Captain Nolan, and Charlotte Rampling, playing Rachel the sympathetic marine biologist, were transformed into figures of myth.

A few days later, Paul Spong found me standing on the boardwalk and immediately put me to work loading wood onto his aluminum-hulled herring skiff. He was barefoot, short and lean, with weathered skin and deeply set eyes. His hairline was receded, but shaggy hair went to his shoulders in back. He was wearing a tan corduroy shirt and baggy, brown corduroy pants with lots of pockets. The skiff had a small, square cabin of plywood and fiberglass. We stacked the wood, which was leftover from an old sawmill on the boardwalk, on top of the cabin and across the bow. "The lowest grade," said Paul, in a lilting New Zealand accent. "A good argument could be made that I should be paid to haul it away." After about half an hour of hauling and piling, he started the 150-horsepower outboard motor and we set off across the strait for Hanson Island. The skiff was sturdy, but it had a tendency to sail in the wind and it banged on the rough water like a garbage can lid. Thirty minutes later we rounded the south end of the island and passed through the channel into Blackfish Sound, on the east side.

A few minutes later we approached Orcalab, a complex of wooden buildings, weathered gray, in a small rock-bound cove, Paul's home for more than twenty years. He had started in a tent. Now, he has a compact, octagonal-shaped house connected by a walkway to an observatory, facing the sound. Another house, two stories tall that he had built for his son, Yashi, was also used as a guest house. The house and the observatory were completed in 1979 and occupy a special use permit, good until the year 2004. The rest of Hanson Island is controlled by Fletcher Challenge, a timber company based in Auckland, New Zealand, and the second-largest forest products company in British Columbia. Logging roads were cut through the island's forest in 1983.

The tide was low when we arrived and the sky overcast. We tied up the boat and unloaded the wood on beds of slippery, brown kelp. My hiking boots were useless on wet rocks and kelp. I went sprawling several times and understood why Paul was barefoot. A pile of logs, bleached white as bones, was washed up beneath the observatory. The wood planks of the deck, the walkway to the observatory, the trail from one house to another, all were soft from the rain. It was quiet, except for the hum of a generator.

Paul had me stash my rucksack in Yashi's house, who was away studying in a film program at Simon Fraser University, then told me I'd have to fend for myself while he took care of business. He disappeared with a wave of his hand. I went for a walk in the surrounding forest, which was dripping from rain and mushy with moss. For ten minutes, I followed a trail into the woods that brought me to an enormous cedar tree, about twelve feet in diameter, with hues of rust and gray and green. Twisted, thick, and ropelike, it looked as if it had spiraled as it grew high above anything nearby. I was told it was about a thousand years old. About thirty feet up was a burl, like a watermelon-sized goiter on the side of the tree. I touched the tree. It was wet, soft, and massive and smelled ripe with decay. There was a small plaque—"Blessed Be He Who Leaves These Trees"—which must have been left to stave off the timber company. In 1975, a special use permit until the year 2004 must have seemed like a grant that would last a lifetime. Now, it seems it will run a bit short of that.

I followed the rough trail through the forest and up the backside of a granite bluff that overlooked the sound. The ground squished and every fifth step I found a five-inch banana slug creeping along at the pace of a glacier. The fir trees were tall and thin and were draped with a fine moss called "witches beard." When I reached the bluff, I could see the observatory below, about as big as my thumbnail. I sat, wet from the walk. Steam rose from the rocks. The clouds were thick and pulsed in long curls through the ravines on the fir-covered hillsides. But then, as if I had reached up and punched a hole with my fist, the sun poured through the clouds at a single spot, just for me. I removed my shirt. After days of sleeping in the rain, the sun felt so warm and sensual, it was difficult to believe I was a warm-blooded animal. I felt like a lizard, soaking up sun on a rock. It lasted only five minutes, then the hole closed, but it was worth a week of sunshine.

The cove and sound below were empty, save for a single,

small boat. It was so quiet, I heard the man in the boat cough, and when he finally started his outboard motor and began puttering away, the noise seemed offensive. I had just said to myself "prime eagle habitat"—open to the sky, with thin, tall, and rather bare fir trees—when a bald eagle flew by at eye level, just a few feet from my nose. I instinctively ducked. The eagle was oblivious to me, flapping slowly and regularly—"whoosh, whoosh, whoosh"—headed for something, somewhere, and as it rose, it was joined by two others. They rose higher and higher in magnificent circles, then fell into a long, slow glide that may have lasted five minutes and finally took them over a ridge to the far side of the island. I heard another engine puttering below. Paul zipped by in his herring skiff, headed for Alert Bay, about forty minutes to the northwest on Cormorant Island, where he shops for groceries and receives faxes. He waved as he passed below the bluff. The skiff bounced. He passed a tugboat chugging in the opposite direction, which was towing a barge with ten thousand tons of logs, moving through the sound as confidently and implacably as an orca.

         Paul went to work at the Vancouver Aquarium in the fall of 1967 after completing his Ph.D. at UCLA in physiological psychology, with a thesis on the relationship between brain mechanisms and behavior. He was also granted a postdoctoral fellowship to conduct a computer analysis of brain wave patterns. He was hired as a researcher at the University of British Columbia—official title, assistant professor of psychiatry—under an arrangement by which half his salary would be paid by the university and half by the aquarium. He looked at it as a simple job opportunity. He was interested in brain-behavior relationships, and it was an opportunity to work with a big-brained animal, Skana, the orca at the aquarium. In absolute size, the brains of many whales are far greater than those of humans. The human brain weighs anywhere from 2.8 to 3.7 pounds. The brain of the orca weighs approximately 14 pounds. The brain of a bottlenose dolphin has been measured at 3.1 pounds. Even though comparing the absolute size of the brains of animals from different species is no more an indicator of anything than the head size of different human races, that brain must be used for something.

In preparation for his first job, Spong had read, among others, the work of John Lilly, who in 1961 wrote *Man and Dolphin*, with its thesis that eventually man would be able to talk to other species and his theories about how that might be done. Toothed whales, he said, surpassed humans in three important areas of measurement of intelligence—brain volume, brain convolutions (the folds in the surface of the brain), and social interactions among individuals. Since whales and dolphins lack hands, Lilly speculated that their evolution might have taken the path of legends and verbal traditions, rather than that of written records and, being sonic creatures, that perhaps their brain functioned as giant sound computers. Lilly's theories and books, however, while they struck a chord with the general public, were generally scorned by most scientists as being the wildest of speculation on the scantiest of scientific data.

Lilly's study of dolphins in laboratories in Miami, Florida, and the Virgin Islands were attempts to train dolphins to speak or mimic a few words of English. Other researchers at the time, though, such as Gregory Bateson, a leading British biologist, anthropologist, and geneticist, discounted the possibility that their vocalizations might be comparable in structure to human language. Bateson, best known in the United States as the former husband of Margaret Mead, wrote, "I personally do not believe that dolphins have anything that a human linguist would call a 'language.' I do not think that any mammal without hands would be stupid enough to arrive at such an outlandish mode of communication." Bateson thought that, through evolution, all expressive devices in dolphins had given way to whistles and creaks. "It is reasonable to suppose that in these animals' vocalization has taken over the communicative functions that most animals perform by facial expression, wagging tails, clenched fists, supinated hands (rotation of forearm so palm is up), flaring nostrils, and the like." Dolphins appeared to communicate using their voices, but their tone indicated they might possess a "digital" language whose primary content was social relationships. "We shall not know much about dolphin communication until we know what one dolphin can read in another's use, direction, volume, and pitch of echolocation." It may be, said Bateson, a system "we terrestrial mammals cannot imagine and for which we have no empathy." He suspected dolphin messages might resemble music more than English.

As Sterling Bunnell explained it in the popular book, *Mind in the Waters*, our primary sense is sight, which provides complex information in the form of pictures and is oriented toward exploring space and distance, but which has poor time discrimination. Our languages are comprised of fairly simple sounds arranged in elaborate sequences. The dolphin and orca auditory system—the primary sense—is spatial, like our eyesight, with much simultaneous information and poor time resolution. So dolphin "language" apparently consists of extremely complex sounds perceived as a unit. A whole paragraph worth of information might be conveyed instantaneously, perhaps, in a certain sense, similar to written Chinese characters, in which a single character contains a package of information. Since their echolocation system gives them detailed images of objects, they might even be able to recreate those images in their speech and thus directly project them to one another.

While familiar with that work and speculation, Paul had also spoken with Kenneth Norris, then at UCLA, today at the Center for Coastal Marine Studies at the University of California, Santa Cruz, and one of the world's experts on whales and dolphins. He advised Paul to study the orca's vision, a limited idea that lent itself to the scientific method. While many researchers were intent on talking to whales and dolphins, the simplest physiology was not understood. Paul took Dr. Norris's advice and did study the orca's vision using simple visual discrimination tests.

During those tests, however, in the holding tank at the aquarium, which was not yet a killer whale habitat, Paul had his lack of expectations challenged. In daily sessions of seventy-two trials, Skana never scored below 90 percent on the visual discrimination test. One day, from one trial to another, the score suddenly dropped to zero and stayed there. When Paul returned the following day, again the score was zero. Four days later, the score was still zero and, reluctantly, Paul began to believe that the zero score was intentional.

One early morning, in late August of that year, he was sitting on a small platform at the edge of the pool. He was dangling his feet slightly above the water. Skana swam in circles below. Suddenly, Skana surfaced and dragged her teeth across Paul's feet, touching the soles and the tops at the same time. Stunned, Paul jerked his feet away. A minute or so passed as he contemplated a life without feet, but he regained his composure and returned to the side of the pool. Again, he dangled his feet above the water and, again, Skana circled the pool, surfaced, and dragged her teeth across Paul's feet. He jerked his feet away a second time. That scene was repeated eleven times, and by the twelfth, Paul's curiosity had overcome his fear. When Skana dragged her teeth across his feet, he left them dangling over the water. Skana stopped. Paul felt as if the subject had turned experimenter.

After that, Paul began to change his experimental methods. He provided Skana with new sounds—bells, a guitar, his flute, even a hi-fi system, with an underwater speaker sealed in a tin can. He found that Skana would perform tricks just to receive new sounds and he concluded that the whale had been "acoustically deprived." While not part of an experiment, he rode around on her back standing up. He continued with this change in method until the following spring. In June, he was delivering a lecture at the University of British Columbia, dealing, for the most part, with his vision experiments, about which he had already published several scientific papers. He concluded that the orca could see in the water about as well as a cat could see in the air, although, of course, its vision would be limited by the clarity of the water. In his summary presentation, though, he revealed some of his other observations: Skana could reason; she understood who she was and who her family was; and the conditions she was being kept in were impoverished. She was, he said, "starved for stimulation." He recommended that the killer whale be put back into the ocean as quickly as possible.

Unfortunately for Paul, then as now, whales make headlines. A reporter from the *Vancouver Sun* was in the audience. The following day, its page-one headline read, "Friend Wants Skana Freed." What could have been an in-house debate quickly deteriorated into a public spitting contest. Naturally, Dr. Newman, director of the aquarium, didn't agree. He thought that if anyone was having a stimulation problem, it was probably Paul, not the whales; and even if the whale did have a problem, it could best be addressed by providing company, not by emptying the tanks. Paul was quickly portrayed as a bearded, long-haired scientist who had taken a flight of fancy and never returned. He looked the part and, having already been told his research days at the aquarium were over, he didn't help his case by continuing his flight of fancy in the newspaper. In an interview with the *Sun*, given under a tree outside the aquarium, with an aquarium public relations person standing nearby, he said that the controversy had spoiled his plans to swim with Skana in the nude, which was a joke, since the water in the aquarium pool was much too cold and Paul always wore a diving suit. He added, "I've always liked Dr. Newman, but he's just a bit dull." The following day, the *Sun* headlined: "Skana Loses Spong the Skinnydipper." Paul's contract with the aquarium, which was due to expire at the end of the month, was not renewed by the dull Dr. Newman. The aquarium spokesperson, asked for a comment, said, "I am told Dr. Spong is not fired, but his contract has not been renewed."

A few days later, in the final newspaper articles on Spong's aquarium departure, Dr. Newman said Spong's statement that Skana would be better off set free had nothing to do with the situation, that, in fact, Spong's experiments had been potentially dangerous for the whale. The bells and glasses he dropped into the tank could have been swallowed and might have killed her. In his defense, Paul, asked to comment on Dr. Newman's accusations, could have argued any number of things. Like Paul Feyerbend, a philosopher of science, he could have argued that his methods were not unscientific, that science, although it seems to have become a modern religion, has no better claim on our support than voodoo. In scientific inquiry, whatever method is proposed, he argues, there will always be circumstances in which that method is not the best one to follow. Paul could have said that scientists live by a calculus of probability that the world tomorrow will be sufficiently like what it is today for the experiments to be completed, but that such faith is grounded in little more than the feeling that the world must have some order to it. He could have quoted Bertrand Russell, who said that such a belief is a bit like the chicken in the barnyard who each day gobbles up the corn tossed to him by the farmer and believes that it will happen each day, until one day the farmer enters the barnyard and rings its neck.

Did he argue any of these things? No. Instead, he said he was creating an organization called "Legion of Orcinus Orca Friends (LOOF)." His last words on the subject, as quoted in the *Sun*, were, "LOOF. LOOF. LOOF. LOOF."

Asked about that, Paul said, "I think it's fair to say that no one has calmed down since." One might think that they should have after more than twenty years, but when I had asked about Spong at the Vancouver Aquarium, I collected an almost unlimited supply of anonymous statements, a practice I thoroughly dislike. I will pass along a few of them, because Paul complained to me that accusations about him float around like some kind of ether and nobody will say anything to him publicly or in person. If for no other reason than that, here are a few comments: "disgruntled former employee"; "a solitary man working with a solitary animal . . . wasn't able to cope any longer"; and, perhaps the ultimate condemnation, "he's *not* a scientist—he says he wants to do this 'benign research' and everybody just buys it, but he hasn't published anything in twenty years." Finally, he was called one of those "people without credentials" who had credibility in the newspaper and television, but had not an ounce of credibility among scientists. Paul's studies of vision were reputable, but his feeling that the whales were sensitive and intelligent, and therefore deprived, was dismissed as "anecdotal."

Of course, there's reason for a certain amount of testiness, since, after leaving the aquarium, Paul continued to campaign for the release of captive whales. For several years, he was heavily involved in the "Save the Whales" campaign for Greenpeace. In 1978, he drove an inflatable boat up the stern slip of a Japanese factory ship, delivered a personal written message to the crew, and handed out Save the Whales buttons. In one sense, his campaign to free the whales was successful. The result of the publicity was that it was practically impossible for the aquarium to capture an orca from British Columbia waters. However, no orcas were ever released. Skana, who started the whole imbroglio, died in captivity in October 1980, after thirteen and a half years of captivity. Paul continued to argue that Hyak, who was a northern resident, should be released to his pod, and in a 1991 *BBC Wildlife Magazine* he wrote that such a release could be an experiment in "family reunification." Time was of the essence, he wrote, since the whale wouldn't live forever. One month before the article was published, Hyak died.

In the study of animals, once a picture has been formed, its truth is only given up reluctantly. Donald Griffin, for example, an ethologist and author of *Animal Thinking, Animal Awareness*, has described setting up an American Association for the Advancement of Science meeting in 1948 for Karl von Frisch, who summarized his experiments on the dances of honeybees. His own reaction to von Frisch's work, he writes, was "frankly incredulous," even though von Frisch's work on color vision in bees and the hearing of fishes was highly regarded. Von Frisch's discoveries that honeybees communicate symbolically "shook up my thinking about the capabilities of animals," said Griffin, and while his own work remained directed toward a mechanistic explanation of animal behavior, he began to believe that the mechanisms must be "much more subtle and versatile than I had imagined."

Griffin, at the time, was working on the echolocation of bats. It was thought that echolocation was only a method of detecting stationary obstacles and avoiding collisions; small insects, it was thought, would not return strong enough echoes to be audible and, thus, rapid and intricate maneuvers must be guided by vision. Griffin found, instead, that echolocation was used in the hunting of small, rapidly moving insect prey. Beginning in the mid-1970s, Griffin writes, he underwent a further change in scientific outlook. Slowly, over the years, he became more and more dissatisfied with the reductionist and behavioristic viewpoint in biology and psychology. In particular, he began to doubt the wisdom of totally ignoring the possibility that animals might experience conscious thoughts and subjective feelings, which is what finally prompted him to attempt to launch the subdiscipline of cognitive ethology in the 1980s.

Griffin wondered—in writing—why it took him so many years to speak up and suggested it was due to his early indoctrination in science at Harvard and elsewhere in the 1930s.

"Many scientific developments and much shaking up of prior ideas were necessary before I was ready to think seriously about the thoughts and feelings of animals." But, as he found his thinking transformed, Griffin, after a distinguished career as a professor at Cornell, Harvard, and Rockefeller universities and a fellow of the American Academy of Arts and Sciences, has also found himself dismissed as "a sentimental softy" for his views on animal consciousness, as he was in a review of his latest book, *Animal Minds*, in the way one might dismiss the views of a kindly, but daft, uncle.

Unlike Donald Griffin, whose views changed over decades, Paul Spong was apparently shaken in a single moment. He was then rejected as not being a scientist, as being unable to cope, because he questioned the notion that, in the name of science, it is right and good to capture and study anything we please, that the knowledge we gain in doing so justifies the inconvenience or pain of the animal. Paul had been hired at the aquarium to study vision. He had changed the question. He then found that, in some situations, principles of logic and organization of data are not just instruments and tools; they are the only kind of thinking permitted. Yet if we applied such a standard to our daily life, we would be unable to get through the day. Most of what we know in the arts or the law, for example, is not based on controlled studies, and that does not mean the arts and the law are without value. There are many different ways of knowing. There is no single, infallible form of knowledge against which all others must be measured. The knowledge we gather is, at best, a snapshot, like the snapshot of a dorsal fin, a piece, a fragment of the truth about the orca. Sometimes, we realize that a great deal of the truth remains hidden from us; many other times, we insist that our snapshot is the truth. But, there could be several snapshots, each taken from a different angle under different conditions, and used at will, when necessary. There is the snapshot of the dorsal fin. Paul provided another snapshot. The child's delight as he or she is drenched by the orca at the aquarium is still another. While the aquarium quickly discarded Paul's snapshot, it considers the snapshot of the child valuable and is prepared to pull it from the wallet on any occasion as evidence of the need for its own existence, although that snapshot is every bit as anecdotal.

         Paul returned in late afternoon, a twenty-foot log in tow behind the skiff. "Sportslogging," he called it. The rain had stopped, but the sky remained slate gray. He now wore two sweaters—a fisherman's knit and a navy blue wool sweater over that—with a navy blue watch cap, but his feet were still bare. After beaching the log, he took me in tow and suggested a chat while he took the skiff across the sound to Parson Island. When we loaded a transmitter on the skiff, I started to wonder just how much of the trip would be a chat and how much of it heavy hauling. The tide had come in, though, one of the biggest of the year, so as I helped Paul with the transmitter, a heavy black box that looked like a stereo speaker, at least I wasn't threatened with falling headlong into the rocks and kelp.

Spong had six remote sites, each with a hydrophone and radio transmitter: two on Hanson Island, called "local left and right"; one on Flower Island, a bit to the north; one directly across the sound on Parson Island; another just to the south on Cracroft Island; and the last on Vancouver Island, south of Telegraph Cove and near the entry to Johnstone Strait. The orcas, as they travel south, usually pass all of those locations. Their calls are picked up by the hydrophone and transmitted to a designated receiver at the base station—Orcalab. Each transmitter also had anywhere from one to three photovoltaic panels to charge the batteries. In spring through fall, the operation is basically maintenance free. In the winter—and the weather during the past few weeks had left people wondering why winter had arrived in July—the solar panels don't work as well. Four of the sites, however, the core network, can be operated the entire year. Paul's goal for more than a decade has been to build a network to study the northern resident orcas year round, without having to capture them or chase them in boats.

On Parson Island, Paul tied the skiff up to a thin fir tree and we hauled the transmitter up above the rocks. The transmitter was being replaced because the Forest Service was picking up fish sounds on the emergency forest fire frequency. The Forest Service had been very nice about the whole thing, said Paul. He busied himself connecting the cable and securing the transmitter. I sat on a wet rock and asked him about his departure from the Vancouver Aquarium.

"I came from a pretty solid academic background—solidly credentialed you could say—so I couldn't be attacked on credentials," he said. "The result was that the whole thing got out of control. But there was no getting around my basic point. I've developed a philosophy that pretty much says leave the whales alone. I'm aware that my attitude has generated a lot of hostility," he laughed at his understatement, "but still my opinion is, everyone will come around in the end, that I'm right, you see. They are too big, too beautiful, too complicated. Captivity in a concrete pen is simply inappropriate.

"Now, there has been a certain amount of antagonism from the scientific community over this. Down at Santa

Cruz, Ken Norris"—who had originally suggested Spong's vision experiments—"has an absolute fit when you raise the question. It's as if holding that opinion was in violation of some scientific truth, so that even raising the question arouses a tremendous amount of anger. I can understand how aquarium directors and trainers and so on would feel defensive—we're talking about their livelihood, after all—but a scientist? I can't understand a scientist reacting that way."

He paused, concentrating on the transmitter. "I understand control in science," he finally said. "I mean, you have the animal there in front of you and you can put it through the paces and it's easy, it's convenient. At the same time, it's a changed animal. It's a caricature. I think an orca in captivity is only an orca for about two years. After that, who knows what they are, besides a great business. They bring in a lot of cash and, really, the science is a side issue. They talk a lot about science, but it's really directed toward animal husbandry and the hope that they can someday successfully breed them. What they're most interested in is keeping the animal alive and keeping people coming through the turnstiles."

He found a place to sit on the rocks. I was cold. He was impervious to it. Suddenly, an enormous black-hulled cruise ship appeared out of the gloom. Was it the Princess Cruise Line? Was it the love boat? He once felt isolated, Paul said, but now floating hotels pass by, two or more each day during the summer.

"If Murray Newman were here he would say that, without the exposure given to the whale by the aquarium, people wouldn't care about them," I said.

It's an ethical question, he answered, the same one faced when science systematically harasses animals in research.

Even though the benefits of that particular research may be high, at what point do you say, that's enough? "If you make a list of what they are and, therefore, why they should not be captive, you come up with the same kind of reasons that caused us to free the slaves," he said.

"Do you have any regrets about the way you left?" I asked.

He thought for a moment, as the cruise ship faded into the gray of the sound and its wake crashed on the rocks below.

"Sometimes I have twinges," he said. "I realize that at a certain point I blew it." But, he hadn't really left science, he said, he just decided he didn't like many of the scientific methods, methods used because many of the people in science were driven to get a degree or a credential or publish a paper and—even though it was a nice, comfortable world to be in—the end result was a lot of trivial busywork. "Since I left that world, I lost that drive. I don't have to produce papers. I'm curious, interested, and I want to know the truth about these animals, but I don't have those demands to make me busy."

"Well," I said. "I was told, flatly and simply, that you are not a scientist."

"And that astounds me," he said. Again, he laughed. "There seems to be an attitude that science is one thing and only one thing. Is science a single-channel venture? Listen to them now say, Well, we can do science in captivity that you can't do in the wild. *Now* they say that. What a waste of scientific opportunity it's been, when they could have been finding out details that, it's true, you can't find out in the ocean. Their primary purpose is to show animals and I really think that period is coming to an end. It's unfortunate that the period of good, controlled access was not used well. Even today, the idea that an aquarium—the Vancouver Aquarium—is an educational facility is absolute nonsense. It's just such a tragedy, in a way. It was a marvelous opportunity and then I got fired and this whole paranoid state set in, where they couldn't trust people to have access. Now, twenty years later, they're still saying, we're going to do science. I was told they're going to test the frequency range of their hearing. Well, when I was there, we tested their eyesight. Twenty years later they test their hearing? What have they been doing all this time?"

He stopped, laughed, and paused for a few moments. "In 1972, we took some people from Alert Bay to see the whales," he said. "Now, from that handful of people, you have thousands here every year. That was probably the first whale watch." He laughed again. "You know, I saw a line in a federal committee study on whales in Johnstone Strait, it said something like, 'the study of orcas in Johnstone Strait has gone from 'several' people in 1970 to dozens or hundreds or whatever the figure is today, and I thought, who else was up here then that I didn't know about?" He again laughed, genuinely lighthearted. "I mean, they can't even give me that, you know?"

In 1960, the Federal Department of Fisheries tried to reduce the number of killer whales at Campbell River, about halfway up the east coast of Vancouver Island, by mounting a .50-caliber machine gun and opening fire when they approached from the west. Twenty-two years later, in June 1982, the British Columbia government established a 3,084-acre marine ecological reserve at the Robson Bight in Johnstone Strait, the site of—as I heard it described on my small shortwave radio—the "world famous" rubbing beaches, where the orcas that frequent the area often stop in the shallow water and rub themselves on the small rocks. An upland buffer area of 1,250 acres was added in 1988 and 1989. Yet, if the progression from machine gun to reserve indicated peace between the orcas and the fisherman, Paul said he still sees whales with bullet wounds. In the same year, the timber company MacMillan-Bloedel started a new conflict by beginning road construction in the area. MacBlo has a "tree farm license" in the Tsitika River watershed, which empties into the Robson Bight, the last unlogged segment of an old growth valley on the east coast of Vancouver Island. Old growth forests are defined as ecosystems dominated by large conifers such as Sitka spruce, Douglas fir, western red cedar, and western hemlock. They are usually more than 250 years old, sometimes more than a thousand. Ninety percent of the remaining old growth forests on Vancouver Island are scheduled for logging under such tree farm licenses. The province advertises itself as "Super Natural British Columbia," but logs its forests at an annual rate of 600,000 acres, more than is cut in a year from all the national forests in the United States.

In 1990, MacBlo finished its roadbuilding, and on October 18, the day Michael Bigg died, they began logging. Five weeks of civil disobedience followed, more than thirty people were arrested, injunctions were filed in the courts both to stop the civil disobedience and to stop the logging, and the British Columbia government recommended a moratorium on logging in the Tsitika. Environmentalists, who want the logging stopped, have focused on its effect on the orcas, insisting that logging will ruin the bight and drive the whales away. An environmentalist from the Western Canada Wilderness Committee said of Robson Bight—"let's remember that without

doubt it is part of the killer whale heritage . . . ," although
the fossil record is fairly sparse in supplying clues to what a
killer whale heritage might be, except that they were apparently once land mammals that returned to the sea and the
oldest fossil whales are from about sixty million years ago.
MacBlo, for its part, was insisting that a halt to logging
would mean a loss of money, jobs, and furthermore, the watershed could be safely logged without harming the bight.
They, too, as I heard a logging company representative ex-
press it at a public meeting, are "on the side of the whales."

The whale watchers, too, were being looked upon as a possible threat to the whale sanctuary. At least one Ph.D. has
been earned by the study of whale watchers—"Nonconsump-
tive Use and Management of Cetaceans in B.C. Coastal Waters (1988)"—so, already, people are closely watching people
watch whales, which demonstrates just how finely the research pie is being sliced. Whale watchers of all kinds are
now advised by the British Columbia Ministry of Parks to stay
away from the whales in the bight. It is illegal under Federal
Fisheries Regulations to disturb or molest killer whales, with a
maximum five-thousand-dollar fine for a conviction. Jim Borrowman told me he used to be able to take his whale watch
tours to the bight for a walk around. Now, he never goes near
it. While I was in the area a whale watching tour operator
from Port McNeil was charged under the act, after a zodiac
boat full of Italian tourists allegedly buzzed an orca.

Prior to the Second World War,
the sea was considered a "world of silence." As a result of the
research done into underwater sound detection and the development of the hydrophone after the war, that silence was

broken. Although even Aristotle had listed a number of sound-producing fish—the catfish, for example—in *Historia Animalium*, most scientists regarded fish sounds as defense mechanisms and antipredator devices. Apparently, though, the problem was that we are not well adapted to hearing sound waves underwater. The sea was always full of noise. We just concluded it was silent until we could hear it. The end result of that discovery?"

"Listen," Paul said.

He was sitting at the Data Train computer, answering correspondence, a VHF radio nearby and a Bill Reid killer whale print on the wall above him. Helena Symonds, his companion, and their ten-year-old daughter, Anna, were sitting at the table near the kitchen. Both of them are blonde haired, although Helena had a few streaks of gray. I was sitting on the floor of the living area of the house, a three-quarter circle with eight large windows that face the cove, in the direction of winter storms. Rain had been drumming on the roof for twenty-four hours, and the green water in the cove had been whipped to cream. The wood stove was fully stoked. White plastic buckets had been placed on the deck to collect rainwater. The house used gravity-fed water from a spring, although during dry periods it didn't work. The buckets were the backup system. Helena, a schoolteacher in Alert Bay when they met some twelve years ago, was at first resistant to the idea of running water, although now she wouldn't be without it, and Paul insisted that someday they'd have running *hot* water.

Paul wanted me to watch some videos, which had taken some time, since the generator had to be cranked up. In the meantime, Anna made popcorn. One of the videos was from French television. It had gone something like this:

FADE IN ON: a Twin Otter airplane circling and landing with a huge splash into the rough, pewter gray water of Blackfish Sound. The background is thick forest and there isn't another human or animal in sight. The plane taxis across the water.

CUT TO: a close-up of the intrepid Krov Menuhin, son of Yehudi Menuhin, stepping out onto the pontoon of the Twin Otter. He holds his hand aloft and shouts, "Paul," as if he's Henry M. Stanley, having finally found missionary explorer David Livingstone.

CUT TO: Paul Spong, standing on the deck of Orcalab, shouting a return greeting. "Krov," he shouts. He smiles, as if he's calling to a dear friend he hasn't seen in twenty years and the look on his face seems to say, "How in the hell did you ever find me, Krov, you crazy, but courageous, man?"

At that point in the video tape, Paul had looked up from the computer long enough to say, "That was take four."

"Four?"

"They had to shoot that scene four times."

He was interrupted by a high-pitched whine coming from a speaker mounted near the kitchen area and attached by cable to the laboratory.

"Listen," Paul said, a second time.

"Whales," said Helena.

"That's nice," said Anna, in a ten-year-old "I could care less" tone, her mouth full of popcorn. She was busy with a thousand-piece puzzle. Anna was schooled at home—third, fourth, and fifth grade—although the plan is to have her attend high school in Victoria, where she can live with Helena's sister. Earlier, she had asked me, "Did you see the cedar tree?" "Yes," I answered. "Did you see the bluff?" "Yes," I said. She nodded, satisfied. I had seen the sights. Despite the apparent isolation of Orcalab, quite a number of visitors pass through during the summer: researchers; students who want to volunteer; and, each year, at least one group of tourists from Japan, who stay in Alert Bay so they can shower with hot running water. Anna's responsibility was to ensure they see the important attractions.

I followed Paul and Helena to the laboratory. Orcalab had two windows facing the sound; a storage room; a deck with three spotting scopes; recording equipment in the center of a long, wooden table; and boxes of tape. Everywhere I looked, there were Fuji audio tapes, labeled with time, date, and identification of pod. Paul put fresh tapes in the recorders.

We could hear the orcas: single calls, long, high-pitched whines—one close, then one distant—from the Parson Island hydrophone, formerly broadcasting to the Forest Service, now strictly to Orcalab. Then another call, an even tone, with no variation in pitch or volume, a long "beeeeeeeeee," which abruptly stopped. An orca can be heard up to about five miles away.

"G's," said Helena, identifying the pods. "And some A's."

Sound alone was enough. The orcas were somewhere in the gloom of Blackfish Sound, sliding through the changing light and temperatures, calling to one another in whistles and beeps. We heard a few streams of the geiger-counter–like echolocation clicks, then what sounded like a rusty door hinge creaking open.

"A30," Helena said. "I always like that one, it's sort of sweet."

A chart was placed beneath the northernmost window, "Port Harvey to Queen Charlotte Strait #3596," and an old window frame had been placed over it, as a viewfinder. As the whales pass through the sound, times and positions of the whales were marked on the window in grease pencil. Earlier that day the pods had been traveling north—3:49 A.M., near Cracroft Point; at 5:06 A.M., near Parson Light; and 6:08 A.M., off Bold Head. Another chart, on the wall of the shed, explained the photovoltaic-powered remote hydrophone network. A third chart described patterns and trends in the orca population of the Johnstone Strait area, 1985–89. Some of the observations: traditionally, A5-pod had treated the area as home turf and in the summer months could be seen on nearly half the days, but for the past two years they had been seen on only 15 percent of the days in the summer and fall, and that seemed to be part of a five-year declining trend. Orca pods often entered Johnstone Strait via Blackney Passage. While sometimes silent during their approach, they became vocal, if only briefly, at the entrance to Johnstone Strait. These vocalizations had been dubbed "announcement calls," and Paul suspected its purpose was to let those already present know there were some new arrivals.

A poster near the door described the steps toward a "benign, nonintrusive observation of Orcas." First, the present network of remote hydrophone stations would be expanded to an area beyond the "core habitat," perhaps tracking pods as far as thirty miles away. Second, a land-based whale watching park would replace on-the-water whale watching, with sites equipped with telescopes and hydrophones on Vancouver Island, Hanson Island, and West Cracroft Island. Paul has tried to interest Fletcher Challenge in sponsoring the project, to make them "allies in protecting the orcas." Finally, what Paul is calling "The Nature Network" is envisioned for the more distant future, a video link to an underwater camera, perhaps using high-definition television, for "live viewing of orcas in the wild." Paul imagines an oceanarium in Tokyo and a museum in Ottawa simultaneously viewing a kelp forest twenty-five feet below the surface near the shore of Cracroft Island, as an orca swims toward the viewer at top speed in pursuit of a spring salmon. In a written proposal, he pointed out that the BBC had been doing live broadcasts of wildlife, that Boston television did live broadcasts of a peregrine falcon, and at the EPCOT Center in Orlando, Florida, the Living Seas Pavilion is linked to the satellite facilities in the EPCOT Communications Center and can show live video from research projects. His proposal for a demonstration on Cracroft Point suggested three cameras—one underwater, one on the surface, and one aloft in a helium balloon—as well as an underwater habitat for human observers. In his proposal, Spong wrote, "How do you allow millions of people a year to experience an environment when that environment is fragile or cold or too small or too deep for people to go to? The answer is to put the eyes and ears, the sensory apparatus, into the field and allow people to 'experience' the reality without actually being there."

"It's coming soon," Spong said, as we huddled in the observatory. "At that point, places like the Vancouver Aquarium will vanish. They will be irrelevant."

"Why do people want to watch whales?" I asked Paul. The orcas in the sound were still calling to one another: squeaks, whistles, and at times, a sound like a cat fight floating out of the storm.

"I have no idea what it is," he said. "Whales make people feel happy. They smile when they see them. So, you wonder, what is it that people respond to? Maybe it's because they're wild? Maybe people want to see them because they represent something wild, and if you see that they're okay, the wilderness is okay, too. Maybe it's because they live the life we wish we could live? I think it's because they are powerful, the most powerful animal in the waters, but they know how to use their power wisely. They are adapted to their environment and don't abuse it. We're also powerful, but we don't know how to use it. Maybe that's it."

"You could also just say, they eat until they're full," I said. "Then they stop."

"Well, people have been reacting to them a long time. It's not something new. The more you see them, the more you know and the more you want to know. The first time I heard them—just like this—I felt awe. I thought, we're not alone. It wasn't something. It was *somebody*."

We listened for a few minutes. A short grunt was followed by a long "phweeeeep," rising in pitch and ending with a punctuation, like a question mark: one close, several very distant, the same call being thrown back and forth seven times. Then, silence. Powerful thrusts of their tails sent them forward, up to take a breath, then down again, scanning with a beam of sound the way we would search the horizon with our eyes, calling to one another through the darkness of the water, perhaps just calling to stay in touch—I'm here—or to say, salmon ahead. I was eavesdropping on another world. I thought of the preface to the *Kokin'shu*, a tenth-century anthology of Japanese poetry, in which it was written, "We hear the bush warbler singing in the flowers or the voice of the frogs that live in the water and know that among all living creatures there is not one that does not have its song."

Paul stopped the tape. The whales had moved beyond the rough monitoring system into their own dimension, away from the probing ears of our desire to know. "I don't think, really, we'll ever know if they carry on a conversation the way you and I imagine a conversation would take place," he went on. "Only once did I ever imagine that was taking place. Just after we had put the hydrophones in the water one year, a young orca stopped right by a hydrophone, it seemed, and he made a loud call. Then again. And again. About ten times, the same thing. Then way out in the distance you could hear a call being made and the young whale made a different call and swam on. I imagined it saying, What is it? What is it? What is it? and the response being, Don't eat it."

"Is it frustrating, to imagine you might listen all these years and never know?" I asked.

"I don't think it's frustrating. I want to know. We all want to know. But it's enough for me to know they're out there. Every time you listen, every time I listen, I feel that sense of something larger, of something grander, of how tiny we are measured against the progression of time. They'll die and pass away, just as we'll die and pass away, and yet that doesn't keep me from wanting to know."

A few days later, I rode in the skiff with Paul to Alert Bay, a village of twelve hundred, half Canadian, half Indian Reserve, split geographically right down the middle. The rain had stopped, although the sky was still the color of slate and the deep green water banged hard on the skiff during the forty-minute ride. At the dock, a four-foot wooden killer whale hung from a structure of three logs, two vertical and an overhead cross piece. It had been painted red and green and carved in the ornate Kwak-waka'wakw style. "Home of the Killer Whale" had been carved into it in six-inch letters. Later, I would find killer whale carvings on the public trash cans and a four foot by six foot painting of a killer whale on the Liquor Store in red, green, and black.

We tied up the skiff near the Nimpkish Hotel, which had been built in 1920 on the shore of the Nimpkish Indian Reserve, but in 1925 had been loaded on a scow and towed to its present location on the Canadian—white—side of town, because beer parlors were illegal on the reserve. It had seven rooms and a gray metal cage around the registrar's window, on which was posted a "Guard Dog on Duty" sign. I suppose that was expected to make a visitor feel secure, but the effect on me was just the opposite. If the front desk needed all the trappings of a bomb shelter, why was the door to my room so flimsy? Downstairs, the bar overlooked the water and the sign on the dock facing the water and incoming boats read, "Where the Fishermen Gather." It was warm.

Paul arrived, having done his grocery shopping, picked up his faxes, and stashed the groceries in the skiff. "Here's what I envision," he said. "A space in the ocean that would be acoustically attractive to the whales, where they'll come and go as they please, free to leave, free to come and interact with humans. It will be a place for interaction."

Then he leaned closer and lowered his voice. "You asked me if I had any regrets. How can I have any regrets, living where I am, doing what I'm doing? I'm the freest man in the world."

A few minutes later, Paul pushed off from the pier. The sky was clearing. The sun was setting. There were about twenty minutes of daylight left. I watched the skiff bang up and down in the chop until it faded from sight.

The following morning, I walked toward the Nimpkish section of town along a paved road that curved for about a mile along the waterfront. I stopped at the Nimpkish Burial Ground, where one of the great native carvers, Mungo Martin, was buried. Martin had died in 1962, just three years after Bill Reid had met him, and the elaborate, thirty-two-foot totem pole that marks his grave, carved by his son-in-law and grandson, was the first pole raised in Alert Bay in nearly forty years. More than a thousand people attended the pole-raising ceremony. It had once been brightly painted in green, red, and yellow. Now, it's weathered and silver with age, showing only a few chips of paint. The top of the pole was a thunderbird figure, but the once outstretched wings had fallen off.

I walked to the far end of the roadway to the U'mista Cultural Center, where I hoped to find Doug Crammer, whom Bill Reid had suggested I ask about killer whale carvings. In particular, he had said, ask him about my killer whale carving. U'mista is a Kwakwaka'wakw word meaning *return*, which was once applied to people who had been taken captive by a raiding party and later had been returned home, either through payment or by a retaliatory raid. Today, the U'mista of the cultural center is the return of native history, language, and culture. Opened in 1980, its only permanent exhibit consists of items once confiscated by the Canadian government when they tried to end the native ceremony of the potlatch, a ceremonial distribution of property used by the natives of the Northwest Coast to validate or reinforce status and to mark certain life passages. Rank and prestige were fundamental to their society. Social position was determined by heredity and wealth. If, for example, two Kwakwaka'wakw had claim to a vacant chieftainship, the one who was able to gather the wealth, with the help of kin, and host the necessary potlatch, could claim the position. Once he had claimed it, however, he couldn't rest. He had to reassert his claim by frequent displays of wealth.

It doesn't sound so different from New York, with its conspicuous displays of wealth and prestige, from the ghastly

Trump Tower or any number of the city's museums, the Rockefeller Wing at the Metropolitan Museum of Art, for example, or the American Museum of Natural History, built to reflect the wealth and importance of the city itself. Yet, as anthropologist Philip Drucker has pointed out, while wealth is associated with prestige in both societies and both spend a lot of time thinking about how to acquire it, New York prestige rests on accumulation, while native prestige rested not on accumulation alone, but the distribution and even the destruction of wealth. Distribution was the only justification for accumulation. To give away wealth was to be wealthy and at the potlatch wealth was given away.

The potlatch was outlawed in 1885. At various times, the law was considered a dead letter, unenforceable and unjust, only to have enforcement revived shortly thereafter. Some Indians stopped the potlatch, many others did not, modifying it or practicing it in outright defiance of the law. Also, there was some irony in the timing of the law. Before the establishment of Fort Rupert in 1849, the potlatch was relatively unimportant in southern Kwakiutl life (Kwakiutl was the term applied to several of the Native American groups in and around the northern end of Vancouver Island by early settlers and anthropologists, including Franz Boas). Once a trading post was operating in an area, however, a dramatic increase occurred in the size and frequency of potlatches, which became a substitute for warfare. In just one example, four groups of Kwakiutl moved to Fort Rupert in 1849, formed a confederacy, and came to be known as the "Fort Ruperts." Those groups had no precedent for ranking the chiefs of the new confederacy and competitive potlatching developed as a way of establishing that rank. More and more elaborate potlatches were held and bitterness grew, as did the quantity of goods given away. Before 1849, potlatches ranged anywhere from 75 to 287 blankets in size, with the Kwakiutl using native blankets. Thereafter, an 1869 potlatch had nine thousand blankets. In 1895, it was more than thirteen thousand blankets and the native blankets had been replaced by the woolen blankets of the Hudson's Bay Company. In general, the wealthier they became and the more sumptuous the potlatches, the greater the necessity to commission art that could display their enhanced status. The whites, whose arrival intensified the practice, then wanted to outlaw it, as if trying to outlaw the symptoms of their own presence.

A further irony was that, by the time of the law in 1885, the native population of British Columbia was in catastrophic decline—in and around Alert Bay, from eight thousand in 1835 to a mere thirteen hundred in 1903—and anthropologists descended on the British Columbia coast like flies on a carcass, staking their own claims for prestige on collections of artwork, totems, and canoes. Franz Boas made his first trip to the Northwest Coast one year after the potlatch law went into effect, with three goals: to establish himself as a creditable scholar of American Indians, by which he could obtain a position at the American Museum of Natural History, in part because his fiancée lived in New York City; to study native linguistics; and to collect artwork that museums would buy. Boas collected masks, cedar rings, and whistles from the Newitti and eventually sold them to a museum in Berlin. During his second trip, in the summer of 1888, the first of five trips sponsored by the British Association for the Advancement of Science, he was asked to do a general ethnographic summary of the area, which included paying special attention to their linguistics and physical anthropology. At the time, that included measurement of head size, since the so-called "cephalic index" was fundamental to the classification of races. Daniel Brinton, one-time president of the American Association for the Advancement of Science, and professor of American linguistics and archeology at the University of Pennsylvania, believed in a hierarchy as follows: white, Asian, American Indian, Australian, Polynesian, and African, each of which had physical traits reflecting their position on the hierarchy.

Boas would later write *The Mind of Primitive Man* and dispel each of the premises upon which some physical anthropologists based that arrangement of races. He would also argue against the potlatch law. In the meantime, he was looking for a job and busy collecting totems, masks, skulls, and skeletons to help get one. In a letter, he described the skulls he had stolen and added, not only did they have scientific value, "these skeletons are worth money." A skull would fetch five dollars, a complete skeleton twenty dollars.

During that trip, Boas met his most valued and trusted informant, George Hunt, who had been born in Fort Rupert in 1854 and raised as a Kwakiutl. Hunt had already worked for Israel Powell, the British Columbia Indian Affairs commissioner, helped collect for the Berlin Royal Ethnographic Museum, and would later collaborate with Boas for the 1893 Chicago Columbian Exposition. Hunt, in fact, was one of the Kwakiutl who spent the summer in Chicago on display. During that time, Boas trained Hunt in ethnographic field methods and later would become infuriated if other museums tried to use him. He told Hunt not to do anything for the anthropologists from Chicago, when they invited him to help arrange the Field Museum's collections, and Boas considered the attempt an "interference with my work." Boas frequently complained of others' vanity, sensitivity, and downright uselessness, and generally thought, as he said in a letter to his wife, "that everyone who is serious should seek and take my advice and not just run out here at random." He even described Hunt at one time as "too lazy to think."

Hunt, if too lazy to think, was not too lazy to collect thousands of pages of texts and purchase approximately twenty-five hundred pieces of art, many of which are the American Museum's most precious examples of Northwest Coast Indian art, including a Nootka Whalers Shrine from Friendly Cove on Vancouver Island, which he acquired in 1904. Of all the Northwest Coast tribes, only the Nootka, on the southwest coast of Vancouver Island, were known to have hunted whales, although others may have at one time or another hunted whales and simply given it up. Philip Drucker, in his book *Indians of the Northwest Coast*, points out that the Nootka actually practiced two kinds of whaling: one, actual whale hunting with harpoons and floats, identical in technique to that of the Eskimo of the Bering Sea and parts of the Arctic coasts; and the other a ritual procedure. Drucker believes the absence of the whaling complex among the other area tribes can only be attributed to its abandonment, because it is inconceivable that those people, so closely related culturally and linguistically to the Nootka, should not have practiced that art at one time. The shrine collected by Hunt is a complex of figures erected in an open shed. Certain individuals, usually chiefs, prayed and purified themselves in the shed to attract beached whales. Two chiefs eventually sold it to Hunt for a total of five hundred dollars, but although described as one of the American Museum's most treasured collections, it was never erected and remains in storage.

While Boas and many others roamed the coast collecting,

the potlatch law became a symbol of the Canadian treatment of the Native Americans in British Columbia. By 1918,
the government had begun successful prosecutions for potlatching. In defiance of those prosecutions, in late 1921 a
large potlatch was held by Dan Crammer—Doug Crammer's
father—on Village Island, the largest potlatch ever held on
the central coast. Among other things, Crammer gave away
twenty-four canoes, three hundred oak trunks, a thousand
sacks of flour, three pool tables, and more than thirty thousand blankets. Afterward, more than fifty people were
charged with having violated the Indian Act and twenty-two
men and women were sent to Oakalla Prison, with sentences
of two to six months. More than 450 potlatch items were
confiscated, including dozens of carved masks.

At about the same time, the American Museum in New
York discovered it did not have enough totem poles to complete its Hall of Northwest Coast Indians. Although Boas
had by then left the museum, the museum called on its informant George Hunt. On June 23, 1923, Hunt wrote to them,
suggesting that Arthur Shaughnessy might be willing to
make four new ones for the museum, but it would have to
wait until he was out of jail. He had been arrested for partici-
pating in a potlatch. "I don't know what he will say after he
comes out of jail for I see that all these poor Indians are now
frightened to do anything . . . yet the government is buying
there [sic] totem poles." After his release, Shaughnessy
carved the poles. Hunt touched them up—he didn't like the
hands—and they are there today, the first four poles as one
enters the hall.

After those arrests and confiscations, the British Columbia government felt that the potlatch had been stamped out.
Yet, a few years later it was again being evasively practiced—

held during Christmas, for example, with the gifts wrapped as presents—or held at distant, inaccessible locations. Attempts at prosecution ceased. In 1951, the law was dropped from the statutes and in the 1960s there was a revival of Native Indian culture in British Columbia. In 1980, the cultural center received half of the confiscated potlatch items that had been held at the National Museum of Man in Ottawa, now the Canadian Museum of Civilization, the other half going to a Museum in Cape Mudge. In 1987, the final group of objects, held by the Royal Ontario Museum in Toronto, was returned.

The U'mista cultural center was a low building on the shoreline, built of cedar planks and resembling a large traditional native house. It had seven-foot wooden sliding doors, which had been carved by Doug Crammer with the figures of Thunderbird and Whale, the Bear and the Wren, the Wild Woman of the Woods, and the Mosquito and Humans who became mosquitos. Crammer had milled the planks, adzed the posts and beams for the big house, and painted the house front design, which faced the water. I was told Crammer came by every day for coffee in late afternoon and that was the only way to get in touch with him, an arrangement that was beginning to make sense to me. I waited. During the afternoon, three people stopped to ask me what I was waiting for. When I told them, each said something to the effect that Crammer was "still bitter," and "he resents whites" and several other similar comments.

Having felt that the objects had been locked up in storage long enough, the Kwakwaka'wakw placed the entire potlatch collection in an open display in a large wooden room with a cedar plank floor and three-foot-diameter cedar beams supporting the roof. My first impression was the warm, sweet smell of the cedar, which reminded me of Bill Reid's studio. To live in a traditional Northwest Coast house was to live inside a cedar chest. I could hear waves pounding on the shore as I walked through the exhibit. Several two-foot-by-three-foot reproductions of letters were also on display, used to explain events and even the conflicting feelings of the period. For example, in one letter, dated December 27, 1918, W. M. Halliday, the Indian agent at Alert Bay from 1906 to 1932, wrote, "I have given the matter of the potlatch a great deal of consideration as it is a mixture of good and evil. One of the chief characteristics of the Indians on the B.C. coast is their hospitality and I have never known of a single instance where anyone was allowed to go hungry while an Indian had food near him. This hospitality has been regardless of race or colour and this quality is cultivated by the potlatch."

On the other hand, in another letter, dated February 7, 1919, Duncan C. Scott, deputy superintendent general for Indian affairs, and at the time also considered one of Canada's finest poets, stated that, "Whatever purpose or principal [sic] of the potlatch may be, the fact remains that potlatches are attended by prolonged idleness and waste of time, by ill-advised and wanton giving away of property and by immorality. The Indians may be able to point to particular occasions on which some of these features did not prevail, but it cannot be denied that these are the usual consequences. . . . The efforts of the Department have been directed to the promotion among the Indians of industry, progress and morality, all of which are greatly hindered by indulgence in the potlatch." And so the battle was joined.

The first group of objects displayed were the "coppers," a round piece of copper pounded into a particular shape. For the Kwakwaka'wakw, the copper symbolized wealth. Each copper had its own story and its own value, which increased every time it changed hands. On occasion, a chief would break a copper, to indicate he was so wealthy he could afford to damage such a valuable object. But, as an indication of the gulf of misunderstanding between white and native, when the whites compensated the natives for the confiscated potlatch items, they judged the coppers to have no value at all. The Kwakwaka'wakw now estimate their value, at the time of confiscation, as thirty-five thousand dollars. Then, came the masks—ravens, bear, eagle, wolf, some mythical birds, and other supernatural creatures from the legends of the Kwakwaka'wakw.

I spent a few hours looking. After the potlatch items, there was a display describing a visit to Alert Bay by the Ainu, the indigenous people of Hokkaido, Japan's northernmost island, today largely dispossessed and basically considered barbarians by the Japanese. There were also a number of paintings and prints from more contemporary artists. A killer whale silkscreen print, done by Doug Crammer in 1987, was on display. It was free and fluid and all motion, a royal blue, black, and red abstraction, with killer whale fins and eyes surfacing and sounding and a black line of U-forms behind them.

Crammer arrived on schedule, in late afternoon, wearing a black sweatshirt, gold rimmed glasses, white painters' pants, and sneakers without socks. He grabbed a cup of black coffee and, with a curt nod, took me into the cultural center's back office, where we sat surrounded by posters for Native American art festivals. Sixty-five years old, Crammer was born in Alert Bay and remembers, as a youngster, watching Arthur Shaughnessy at work. Later, he received his first formal artistic training from Mungo Martin. In the mid-1950s, he met Bill Reid and agreed to work with him at the University of British Columbia on a carving project, which included seven large carvings and two houses, and since then he's made his living as a carver.

He removed his worn, beige jungle hat. His hair was white and cut to the nubs. His dark skin was deeply creased, with laugh lines around his eyes, and he had tough, wiry arms. He quickly lit the first of many Player's cigarettes, and after asking me why I was waiting for him, he nodded and said nothing, so I asked him about the killer whale print. Was the killer whale used because it was culturally important or was he just manipulating forms?

He took a deliberate drag on his cigarette, a sip of coffee, and said, "I wanted to come up with something that didn't look like anything else. I did forty-eight drawings for that and most of them started out as a doodle, playing with this or that form, some of them started out to be something else altogether than what they ended up. In any case, I finished it and I sold it. What else is there?"

"Well, there might be a sense of tradition."

"Anyone who tells you that, it's baloney." He took another long drag on his cigarette and another sip of coffee. "I was able to quit working to do this and I could do that because it sold, not because it carried on a tradition. I worked at logging for eighteen years and also fishing, because everyone here fished from the time they were big enough to pull a net. I quit that work to do this, in 1958 when I started working with Bill. I did some carving before that, but nothing much, and it was successful right away. If Bill had said, come to work for three months, I probably would have stayed a logger, but he said a year, so that made it worthwhile. I was with him two years. We opened up a store. Then the big poles happened. I've been doing it ever since."

Another drag, another sip of coffee.

"Well, what do you think people respond to in the art? In that killer whale print, say?" I had been told that Canadian Native Prints in Campbell River had sold a hundred copies of his killer whale print.

"I haven't any idea why they buy them," he said. "They'll always sell though. They buy one piece and before you know it they have a whole collection." He smiled, puffed, and sipped. "As long as there are white folks around, they'll sell. And I don't really care who buys them. Now, though, they just buy the little signature at the bottom. It's not the art anymore."

"Is that the only criteria for this art, then, that it sells?"

"Well, art, to me, I always thought it had to be original for it to be art." He stamped out his cigarette and lit another. "All this Indian junk is nothing but copies. They'll copy anything—poles, anything. Really, I'm telling you people will buy anything I make. Then, I haven't been doing too much the past few years. I've been fiddling around with houses, the old-type houses."

"What did it mean to get these artifacts back?"

"I don't know why they wanted to get them back, to tell you the truth. It's playing white man. We never had things in museums." He was silent for a minute, sipped his coffee, then leaned across the table, as if he wanted to ensure I could hear him. "And I'll tell you what else. In town these days everyone is raising a stink about some of those poles. 'They don't look pretty anymore. Why don't you put those wings back on the Mungo Martin pole?'" He leaned back and paused. "And it may be done someday, but if it is, it won't be for themselves and it won't be for Mungo Martin. It will be for the tourists.

In the poles, the point is in the carving and the erection. After that, it doesn't really matter if it falls down the next day. The point is in the doing."

"Well, what do you think it is people respond to?" I asked. "People who might buy your killer whale, is the whale something besides an abstract painting?"

"Yeah, it is," and he smiled again, crinkled his eyes, and paused to give his words the proper effect, "To white folks. To us, they've always been there. To me, it's the same thing as having people come to the coast from the prairie and start yelling about how the forest is being raped. Save the whales. Save the trees. Save this and that. I don't think they want to save anything. They just need a cause to be happy. I'm not quite sure what people are trying to save. Are they really concerned about the forest or do they just want to save the view from their window?"

He put his cigarette out and lit another. He puffed contentedly. "So, I'm talking to you because Bill sent you, but are you almost finished?" he asked.

"One more question," I said. "Bill said I should ask you about his killer whale bronze."

He sucked on his cigarette and smiled. "Bill's a great guy and a great artist, you know, but everything you do isn't going to be great, no matter what kind of guy you are."

"And?"

He laughed. I laughed. He paused for a sip of coffee, still smiling. "What else is there to think? I told Bill it was ugly."

I was waiting to board the ferry to Port McNeil, from where I would take a bus to Port Hardy, then another ferry in a few days north to Prince Rupert, and from there still another ferry to the Queen Charlotte Islands.

The British Columbia Info Tourist Center was nearby, so I went in and asked the young, blonde-haired teenager behind the counter, "Why is Alert Bay called the 'Home of the Killer Whale?' "

She looked at me as if I'd hit her with a stick. I pressed her a bit, to break the spell. "Is it for advertising or public relations or something?"

"No, no," she said. "It's Indian. The Inuit word for this area means, home of the killer whale, and they used to call it that because the killer whales come right through here and they sometimes beach themselves right down here on the shore."

All of which was complete nonsense. There is no Inuit word for the area, since the Inuit live in the Arctic. In fact, the Nimpkish didn't have a word for this area either, since they lived on the Nimpkish River and were persuaded to move to Alert Bay, which was named after the British survey ship HMS *Alert* after 1870 to provide the labor for the new salmon saltery. As for beaching themselves on the shore, the rubbing beaches were in Johnstone Strait, about six miles away. I returned to the ferry slip. A middle-aged woman in pale green rain gear was in the ticket booth, so I asked her, "Why is Alert Bay the home of the killer whale?" I pointed to the carving overhead, which was swaying back and forth in the breeze. She looked at it and said, "The Chamber of Commerce thought that up a few years ago."

I boarded the ferry, which was carrying only about ten passengers, settled in for the thirty-minute ride to Port Mc-Neil, and was quickly run over by a five-year-old struggling to secure a better view. His father, Sam Cook, followed hard on his heels with apologies and we slid into a conversation. He was a Nimpkish Indian, with deep brown skin, a faint goatee and mustache, and a quick smile. He told me he had grown up in Vancouver, where he wasn't doing anything except being shiftless, so he had moved to Alert Bay twelve years before, back to his village, back to his roots, and since then he had made his living as a fisherman. While he was fishing, he had learned to read.

"I never knew reading was so much fun," he said. "Now, I can't get enough of it. I've just been tearing through the philosophers. Then, I tell my kids what I've been reading and sometimes—especially my fifteen-year-old—they think, 'Dad, I think you're not all there.' But I say, just try it. I try things. I listen to your music. You should try listening to mine."

"I'm not sure it applies to listening to rock and roll," I said, "but I think Plato said philosophy begins in the experience of wonder."

"Exactly," he said. "But he's not my favorite. My favorites are Chuang Tzu. The Tao. Zen. Those riddles drove me crazy for a while. Like the one—what is in the space between two thoughts? But now I love to talk about it."

I was surprised, because I had also been reading Chuang Tzu, prompted by my girlfriend, Karen, who had been reading him in a study program in New Mexico. Chuang Tzu was a Chinese philosopher of around 300 B.C. who represented the Taoist strain in Chinese thought and its central question—if it can possibly be summed up—of how man can live in a world dominated by suffering and absurdity. The answer? Free yourself from the world, which is the answer of a Taoist mystic. Just a few days before, while I was huddled in a telephone booth during a rainstorm, Karen had told me a sentence from Chuang Tzu had been awarded the "most impenetrable sentence of the semester." It went like this: "There is a beginning. There is not yet beginning to be a beginning. There is not yet beginning to be a not yet beginning to be a beginning. There is being. There is nonbeing. There is not yet beginning to be nonbeing. There is not yet beginning to be a not yet beginning to be nonbeing. Suddenly there is being and nonbeing. But between this being and nonbeing, I don't really know which is being and which is nonbeing. Now I have said something, but I don't know whether what I have said has really said something or whether it hasn't said something."

Karen believed it was a creation story. I wasn't sure what to make of it. I told Sam I was familiar with Chuang Tzu, the Burton Watson translation, at least, and related the problem of differentiating being from nonbeing.

"Exactly right," he said. He was delighted. "It's a problem. But you also have to realize, nonbeing is not empty, which is what most people think. It's 'full of nothing.' It gives a place to the unknown."

I told him that it sounded as if he'd slipped away from his Nimpkish roots, but he assured me that was not the case.

"It strikes a chord, you know?" he said. "Those riddles, what they told me was, you eat, shit, sleep, and die and, if you can do that, you've found the way to live. See each moment, but don't try to make it something other than it is. And I have a sense of peace about being here. I love to feel the rain in my face. I love to feel the sun on my back. There's nothing like it, to sit on a rock and feel the sun on your back. My kids haven't learned that yet. They sit on a rock and in five minutes they're bored. But when they start to rush me I say, 'Great understanding is broad and unhurried; little understanding is cramped and busy.' That's Chuang Tzu, but it's very Nimpkish."

I asked Sam if he had an unhurried understanding of killer whales. He said I should go talk to Paul Spong. When I explained I already had, he said, "I know Paul. I fish over there.

Show up with a flat of beer and he'll drop everything. He's a good guy. And the killer whales, they are beautiful, beautiful." Sam said that many years ago native hunters would paddle their canoes up beside the killer whale—blackfish, they called them—and then get out and walk on their backs to show how brave they were, but he thinks that was when there were a lot more of them. He believes there used to be thousands, that there were probably as many blackfish as there were Nimpkish, and now there are only a few hundred of each.

To the Northwest Coast natives, the killer whale—like almost all animals—is believed to be an ancestor who, according to tradition, had appeared to some ancestor or, in some instances, had transformed itself to human form and become an ancestor, which is why the animals are depicted in the art, on totem poles and houses. The descendants of that ancestor, in the female line, inherit the right to display symbols of the supernatural being, to demonstrate their noble descent. Whether painted or carved, the motifs are often referred to as crests.

"To us, they're people," Sam said. "All animals can transform themselves to human form. And, they aren't so far away from us. They're very much like us. It has enough food, so it has plenty of time to play, and it's a very good life. It's a sensual life. I think that's the way the Nimpkish used to live. Food was always very plentiful, so we had time to play. We carved totems. We danced. We sang. We respected the killer whale because we know that we all come from the same place."

He smiled brightly. "The sage embraces things. Ordinary men discriminate. Those who discriminate fail to see."

"Chuang Tzu?" I asked.

"Exactly right. It strikes a chord. 'The ten thousand things are one with me.'"

The ferry was nearing Port McNeil. Sam rose. He had to go down to his car, a blue Toyota.

"Good luck, man, in finding your whales," he said. He looked out at the light rain in the strait and grinned. "We believe that if you've been a good person, if you've lived right, you'll come back as a killer whale. I hope it happens to me. I'd love it. Swim all over the damn place. Fish all you want. My grandmother used to take me up to the bluff and when the killer whales would go by she would say, 'Yo Gakgump,' wave to your grandfather. So, we'd wave to the killer whales." He looked out at the water and waved his hand, as if he was watching his grandfather pass. "Yo Gakgump."

"Sam, what is in the space between two thoughts?" I asked.

"Peace," he said.

Sam left. Later, while waiting for another ferry, in my Burton Watson translation of Chuang Tzu, I read the following story, which struck a chord:

"Once a seabird alighted in the suburbs of the Lu capital. The Marquis of Lu escorted it to the ancestral temple, where he entertained it, performing the Nine Shao music for it to listen to and presenting it with the meat of the T'ai-lao sacrifice to feast on. But the bird only looked dazed and forlorn, refusing to eat a single slice of meat or drink a cup of wine, and in three days it was dead. This is to try to nourish a bird with what would nourish you instead of what would nourish a bird. If you want to nourish a bird with what nourishes a bird, then you should let it roost in the deep forest, play among the banks and islands, float on the rivers and lakes, eat mudfish and minnows, follow the rest of the flock in flight and rest and live any way it chooses."

QUEEN CHARLOTTE ISLANDS

When pestilence approaches, how will we know it? According to the Haida, the natives of the Queen Charlotte Islands, pestilence sailed in a canoe with huge wings or sails, so when the first European ships were seen off the coast, the Haida would often stand on shore staring at them through cylinders made of skunk cabbage leaves. Only in that way, said their legends, could they look upon pestilence without being struck dead. On a few occasions after I arrived in the Queen Charlottes, I felt as if

the person to whom I was speaking had suddenly screwed a
skunk cabbage leaf up to their eye, usually after I told them
I'd come from New York. The Charlottes, I often heard it
wistfully expressed, are located "west of west," beyond time
and geography, and in Vancouver mention of the islands
would prompt an envious look, as if I were going to the one
place in all of Canada still shrouded in mystery and, therefore, the one place that they, too, had always wanted to go. It
is no accident that the underwater scene in the Vancouver
Aquarium's Pacific Canada exhibit has been designed to resemble the area near Anthony Island, located at the south-
ernmost tip of the some 150 islands. It is intended to take
advantage of that mystique.

In part, the Charlotte's mystery exists because the archipelago, about 150 miles long, is among the most isolated in
Canada, separated from the mainland by thirty to ninety
miles of open water, the shallow and often stormy Hecate
Strait, a six-hour ferry ride from Prince Rupert. The isolation
has prompted some biologists to refer to them as the Canadian Galapagos and an "evolutionary showcase," since the
formation of new animal species is usually dependent upon
such isolation. Some forms of native land mammals and
birds—such as the black bear, the largest in North America,
and the saw whet owl—are unique to the Charlottes and numerous plant species are found only in the Charlottes or in
Japan or Ireland. The presence of such species, however, has
also stirred some debate among scientists. Biologists insist
that the peculiarities of various species must have taken more
than ten thousand years to evolve, while geologists argue that
as recently as ten thousand years ago British Columbia, including the Charlottes, was covered by the glaciers of the last
great ice age. The solution? Another theory, which has pos-

tulated that, although masses of ice scraped the surrounding landscape down to bedrock, plant and animal families survived in ice-free pockets on the islands—"refugia." In addition, while most archaeologists believe people moved into the north coast about nine thousand years ago as the ice sheet retreated, one theory suggests that the refugia were used not only by plant and animal species but also people traveling in boats from one refugia to another, migrating along the coast perhaps as far back as forty thousand years ago.

Before that, long before the first visit of a European—a Spaniard, Juan Perez, in 1774—perhaps a time surrounded by a mist so heavy it seems to be a drizzle, not uncommon in the Charlottes, we find the Raven, the trickster of the Haida tradition, whose antics continually changed and rendered the world, who had always existed and always will exist. It was the time of the great flood that had covered the earth, which had finally receded and revealed the thin strip of land called Rose Spit. There, the Raven found a gigantic clamshell half buried in the sand, full of little squirming creatures. He coaxed them out of the shell. They were strange creatures, pale and naked, waving and fluttering their arms, squabbling and confused—in Haida artist Bill Reid's account, "astounded, embarrassed, and confused by a rush of new emotions and sensations." They were the Haida, the first humans, and they called the place Haida Gwaii.

Those Haida feared the dreaded Ocean people and their fierce chief, the Killer Whale, who was once a man, but went out to fish in his canoe one day and was never seen again. There were some fifty supernatural killer whales, the most powerful of all supernatural people. They lived in towns deep under the ocean, scattered under steep cliffs and projecting rocks, from where they could travel far inland along channels and under the mountains. They could seize men who passed by in their canoes and drag them down into one of their towns, which had been the fate of their great chief. However, the chief never forgot the Haida. He was determined to give them relief from their fear of the monsters.

To do that, he invited the chiefs of the whole earth to a great feast. All the monsters of the Ocean and of the Land and the Upper World soon arrived, using Killer Whales as their canoes. The great chief spoke to the assembly, telling them of the heavy sorrow caused by the Ocean monsters, who killed all the people that passed their doors. In response, there was silence. Then, speaking with one voice, the Ocean monsters promised not to kill people any longer. They kept their promises when they returned to their undersea homes and each one removed his house from the paths of the men's canoes.

Since that day, the great chief's descendants carve the sea monsters on their crest poles in memory of his good deed, which is why, in the mist so heavy to be a drizzle, the Haida had nothing to fear from the Killer Whale.

Etienne Marchand, a Frenchman who visited the Haida in the 1790s, said that what astonished him the most was "to see painting everywhere, everywhere sculpture, among a nation of hunters." It was a highly formalized art, with a long established style, which is still used today. Art embellished most aspects of their lives, from finely carved figures on a halibut hook to the designs on paddles and baby cradles and the totem pole. One art historian and anthropologist, Alexander Alland, has written that every member of every nonliterate culture in the world understands and responds to that culture's public art; the art "re-verberates" across the mythical and the mundane. The Haida art, in particular the totem, carried more than one meaning; it was admired for the skill of the carving and the message it carried. Totems were of three kinds: house frontal poles, mortuary poles, and memorial poles for deceased relatives. The animals used in the totems—the figures of the Killer Whale, the Raven, the Grizzly Bear, the Thunderbird, the Eagle, the Frog, and others—often told a story and explained a family history. They were ancestors and, thus, crest figures the family could claim, which were jealously guarded. In the artwork, the crest animals could be presented realistically or in varying degrees of abstraction, because, as one native artist explained it to anthropologist Erna Gunther, "This is not just the animal, it is his spirit. I see only part of him, but that is him." The artist followed the mythology into the world of imaginary creatures and made them familiar to the people around him. It was a function of belief, of using the materials of their environment—cedar—and an expression of union with the world.

European contact brought an explosion of artistic creation, which some anthropologists attribute to the arrival of metal tools, but that was followed by a rapid decline, since the culture and customs of which the art was so much a part were gradually decimated. In 1786, when the islands were named by English Captain George Dixon after Queen Sophie Charlotte von Mecklenburg-Strelitz, wife of the mad king of England, George III, there were eight thousand or so Haida, with a reputation for being keen traders, fierce fighters, skilled carvers and builders of beautiful canoes. One hundred years later, about six hundred Haida remained, stripped of their customs and ravaged by alcohol and epidemics of smallpox. With sextant and compass, the Spanish and English had obliterated the spaces on their own maps labeled "terra incognita," but had transformed the world of the Haida into a dangerous place, which they no longer recognized. The remaining Haida lived in two villages—two "refugia"—Masset and Skidegate—and my own theory is that the population of supernatural killer whales declined along with the Haida. Cabbage leaves had only briefly delayed the arrival of pestilence.

While the Haida were reduced to a few hundred in Masset and Skidegate, anthropologists arrived, collecting Haida art and earning—some of them, anyway—a reputation for robbing graves. In 1897, Franz Boas sought out Charles Edenshaw and asked him to do several crayon drawings, which were then used in his book, *Primitive Art*, including a drawing of Wasco, a sea monster with a wolf's body wrapping its tail around one skeletal whale and carrying another whale between his long ears. Edenshaw was later commissioned by John Swanton, working for Boas on the Jesup Expedition, to carve some totem poles, a model house, and a model canoe, and make a series of drawings of totem poles and two-dimensional designs. Swanton, a young Ph.D. from Harvard, wrote to Boas that everything was for sale: a totem pole model for ten dollars; totem poles, twenty-five dollars to sixty dollars each; crayon drawings, ten or twenty cents. He also said that C. F. Newcombe, a collector from Victoria, was "dredging the place" before his eyes, which was intended as a mild criticism. Boas, however, didn't mind if a place was dredged, as long as it was being dredged at his direction. Newcombe was soon working for Boas, traveling to abandoned villages and gathering mortuary poles, house posts, and other large totem poles, among them the pole that I had stopped and admired and drawn that hot summer day in New York. Edenshaw, in the meantime, made a model of the house his uncle had built in midcentury in Kiutsa, prior to its abandonment, said to have been conceived in a dream and, thus, called "Myth House." That model is also displayed in New York. Edenshaw, considered one of the master Haida carvers, died in 1920, but thirty years later one of his gold bracelets changed Bill Reid's life forever and today his relative, Jim Hart, who helped Reid sculpt his killer whale bronze, carves wood on the shoreline at the end of a gravel road in the native village of Old Masset. He calls his house "the Ark."

Old Masset is at the northeast corner of Graham Island, the largest of the islands of the Charlottes, near Rose Spit, not far from where human life began. Old Masset, the Haida reserve, is distinct from Masset, which has a population of about sixteen hundred and two Chinese restaurants. I went looking for Jim Hart because Bill Reid told me to find him—he described him as "the real thing"—and I had been told he was carving a killer whale totem. I thought he might talk to me about carvings and killer whales before I headed south by kayak to find another killer whale totem, on Anthony Island in the abandoned village of Ninstints. The Haida totems standing there had been declared a World Heritage Site by the United Nations in 1981. I was also hoping to see the mystery whales. A team of people from the Vancouver Aquarium had established a camp on Burnaby Island, the start of a network to collect observations of those whales.

I found Jim standing on top of an enormous frog, which he had carved in red cedar. His arms were draped over one of the rafters in his workshop and he was peering down at the frog's two bulging eyes, trying to find some balance between them. He would stare intently for as long as a minute, then hop down, use his adz to peel away a few chips from the right eye, and resume his position atop the frog. A faded, late afternoon sun was poking feebly through the four small windows that faced the bay and a cold wind whistled through the cracks in the walls of his studio, a gray and weathered hundred-foot by thirty-foot shack, surrounded by piles of wood, some freshly cut, some soft with age. Jim, tall and slender, was wearing olive green pants and an ivory shirt. His black hair and a single prominent gray hair was tied into a pony tail with a red and black band and reached halfway down his back. He's forty years old. A vague, teenage mustache curled around his mouth.

"As you can see, this isn't a killer whale," Jim said. I had told him I'd expected him to be carving a killer whale and he said he'd done a killer whale totem the year before in Norway, but if I wanted to hang around for a few months he'd see if he couldn't work one up for me. He referred to the sculpture on which he was standing as "frog constellation," because he hadn't yet thought of a name. Off to the side, on a workbench piled high with discarded tools, a roll of toilet paper, and cans of Coke, was the two-foot model, also carved in cedar: two upright frog figures, a male and female, one sitting behind the other, on top of a third squatting frog. The figures were wearing cylindrical hats. He was sculpting the large figure in pieces, which would eventually fit together like a puzzle and stand sixteen feet high in the public atrium of an office building in Orange County, California, although it was about a year overdue. The company that commissioned it was threatening to sue. Jim tried to placate them by sending the model. They returned it, broken, which they blamed on the cleaning lady.

"I'm working on it and it'll be done when it's done," he

said. It was his first law of Haida carving. He jumped down lightly from his perch and continued peeling cedar away from the eyeball, leaning into the adz with his body. He stopped, put on a pair of goggles, and took the adz to one of three grinding wheels on the workbench behind him. Red cedar is soft and requires an especially sharp tool, or it will tear. After a few seconds of sharpening, he turned back to the eyeball and, without removing his goggles, sighted down the adz like a pool shark down a cue. Wood chips curled to the floor.

"Carving with wood makes me feel like I have roots, like I'm planted," he said. "It's solid. The old tradition is solid. There's depth to it. You can feel there's something to it. The old carvings, they were fine, detailed pieces and they didn't have the tools we have. You know, I watch some guys, they just churn the stuff out and it makes me sad." He stopped talking. The chips continued to pile up on the floor. Chip. Chip. Chip. "Or maybe I wish I could do that, churn it out. Maybe that's what's sad. It would pay, that's for sure. But I don't seem to be able to do it. Things could be toppling down around my ears, and I still can't do it."

He continued carving in silence. He was raised in Masset, of a white father originally from Ontario, who died of cancer in 1981, and a Haida mother. The art was all around him when he was growing up, he told me, although he didn't know it. One day he simply realized, "That is art and I should do it," and once he'd decided to do it, he didn't have to flounder. The tradition gave him a way of working and looking. As Jim described it, the art was everywhere and one carver would try a little something new and the next carver would see it and add a touch of his own and it would work like a spiral, just going up and up, more and more and better art, yet all within the tradition.

Jim apprenticed with another Haida artist, Robert Davidson, who had apprenticed with Bill Reid, then lived in Vancouver for several years, helping Reid with his killer whale bronze. He carved the four-foot scale model and helped with carving the Raven and the First Men, before moving back to the Queen Charlottes. Three years ago he began to build his house next to his studio and called it the Ark because he believes one day the tide will come in and he'll simply float away. The design is a traditional Haida longhouse model, with a huge main floor, except it has a basement, which opens to the water, and two Feathercraft kayaks stored in the rafters. When he moved back home, he resolved any questions he had about what he was doing, he said, because his carving connects him with a tradition and the nature of this place.

"This is home," he said, as he continued carving. "Everyone has to make a stand somewhere and this is where it will be for me. Unless I decide to go to New York. I hear there's real money to be made there, eh?" He raised his eyebrows in comment. "I am going to go there someday. I want to see those old poles up close. Although I hear some of them are in a pretty bad neighborhood. You have to run from your car to the door, eh?"

I told him his impression was not even close to being true, although he'd be advised not to leave his radio in the car. I asked him what he thought about the idea that those poles had been collected unfairly; "captured," as it were, or confiscated.

He chipped away for a minute or so, then said, "My idea about that is, we can always make more." He said that the unfinished pole at one end of the shop was destined for a Vancouver home, purchased by a woman as a birthday present for her husband. A pole he had carved last year was now standing in Fort Baker, California, near the Golden Gate Bridge.

The killer whale pole was in Norway. The frog constellation was for Orange County. He looked up and asked, "What do they mean to any of those people?"

"You know that even in the beginning, when the whites first arrived, they couldn't believe Indians did this work," he said. "They were shocked, eh? So, every time another pole goes up somewhere, in Orange County or in San Francisco, that's another way of saying we're still here. So, it will mean something different to the two sides. What it means to me is not what it means to the guy in California. To him, maybe it's a curiosity. Collect 'em up. Have the art without the Indians." He paused and grinned and continued to chip away. "Anyway, I hope they all have a big impact, because that will mean more cash and I can make some more. Besides, how do you put a value on a pole, that's what I always say? I think we played the game with someone else's rules for a long time. Now, we've got to play our game and get as much as we can."

I had seen a fifteen-foot pole for sale at Hill's Indian Crafts in Vancouver for fifteen thousand dollars. I asked, "How much are you getting paid for the pole?"

"Not enough," he replied.

"How about the frogs?"

"Not *nearly* enough."

"Especially if it's a year overdue."

"Well, these follow the wood. They move at their own pace."

The following day in the early morning, a strong, cold wind knifed off the water and the long grass outside the shed was bent sideways. Inside, Jim was sitting on the frog figure's face, using a skew chisel to pare away the wood. It was quiet, except for the whistling of the wind through the cracks. Jim's sleeves were rolled up. His forearms were smoothly muscled and his veins stood out like ropes. He had been at it several hours. He gently brushed his hair away from his eyes, climbed up, and stood on the body of the frog, arms over the rafters for support. He stared down at the face. When I looked at it, I couldn't keep myself from smiling. The frog was full of tension, just barely contained.

"Looks good, eh?" Jim said.

I agreed.

"Long way to go, though."

"How long, do you think?"

"A month and a half."

"Not so bad, then."

"Yeah, but I always say a month and a half. It sounds reasonable."

He climbed down and went back to work, peeling away cedar from the nose.

"So, what did you think of Bill's killer whale?" he asked. "The bronze."

"Well, what I liked was that little wooden sculpture he showed me, not the big bronze."

"I did that."

"You did?"

"Bill roughed it out. I did it."

Quiet again. Chip. Chip. Chip. Falling to the floor softly, like snow.

"When you started asking yourself, why am I doing this, how did you answer?" I asked.

Chip. Chip. Chip.

"I don't know. Look at the Chinese. They would spend generations on a piece of jade. Is that for yourself, from gen-

eration to generation on a small green stone. Why?" He
smiled, not looking up from the face, still leaning into the
chisel. "Tell me, why?"

Off to the side, a technical drawing was tacked to a board,
although it could only barely be identified as such. It was
torn in several places and it was crusted with dirt. I couldn't
read any of the dimensions, so I looked more closely. It was a
drawing of the frog constellation and the tattered look may
have been an indication of just how long he'd been working
on the sculpture. I asked Jim what he intended to do with it.

"Why?" he asked, never lifting his eyes from the frog's.

"If you're going to get rid of it, maybe I could buy it from
you."

"Well, I still need it, although I'm mostly using the model
now. I could sell it to you."

He paused. Chip. Chip. Chip.

"How much?" he asked.

"How much would you want?"

"I don't know. I've never sold a drawing like that before."

"Well, then, maybe you shouldn't sell it. Just don't throw it
away."

"I won't throw it away. I'll just pack it away with my stuff."
Chip. Chip. Chip.

"How much do you think it's worth?" he asked.

"I don't know. I've never offered to buy anything like it."

"It's one of a kind, eh?"

"On the other hand, it's got dirt all over it, it's torn and
it's a technical drawing."

"Yeah, that's true." He paused. "But that could be my re-
tirement you're talking about."

"I think you just soared out of my price range."

"Well, what's fair?"

"You tell me."

The conversation continued that way for several minutes. We never did agree on what was fair. We couldn't even arrive at an opening bid. Two inept bargainers had apparently met their match. We decided to put off our negotiations, perhaps forever. If I had arrived in North America with the Dutch West India Company, as an aide to Peter Minuit, the Algonquin would probably have taken the twenty-four dollars worth of beads and trinkets and cloth, kept Manhattan Island, and charged me rent for a studio-apartment–size plot of land.

Jim stood up and began to sharpen his adz. He smiled. "Now, we should look out the window here and see killer whales going by," he said. "Keep your eyes open."

We both stood and looked out at the bay. Nothing. Only the wind, bending the grass. Inside, Jim smiled and returned to work. Chip. Chip. Chip. He possessed the same impulse that drove his ancestors, whose carvings are now thousands of miles away in New York City, an impulse sending out thin filaments of imagination to entangle us with each other. The frog's eye was bulging. The air was tangy with cedar. Jim carved. He smoothed. Chip. Chip. Chip. He remade the world.

One of the Haida stories about the killer whale goes like this: One day some men were out seal hunting and a killer whale appeared near their canoe. The young men threw stones at the whale and struck its dorsal fin many times. The whale then left the men and headed toward the shore, and soon afterwards, they saw smoke rising from the beach. The hunters approached the beach to investigate. When they reached the shore they came upon a man cooking some food. "Why did you throw stones at my canoe?" he asked. "Go now and get some cedar from the woods and mend my canoe." The hunters mended the canoe, and when they had finished, the man instructed them to turn their backs to the sea and to cover their heads with their robes. "Don't look until I call you," were his instructions. They heard the canoe slide down into the water, then the man called out, "Look now." The hunters turned to see the canoe in the water, but as it reached the second breaker, it plunged beneath the surface. What surfaced beyond was not the canoe, but a killer whale. The man spirit was inside the killer whale.

The killer whales, according to the Haida, possessed the same kind of souls as man and could change into human form at will. In effect, they might as well be thought of as people, with special abilities, their own territories, villages, houses, canoes, and chiefs. In their own houses, the killer whale would use human form. When they wished to appear in their animal form, they put on cloaks and masks and spoke their killer whale language. Similarly, when the Haida put on masks and cloaks during rituals and carefully mimicked the grunts and calls of animals or birds, they tried to enter a mental state that included them in the animal society. The Haida watched and listened to killer whales and all animals closely. While today ethologists studying animal behavior might point out resemblances to human patterns, the Haida studied the animals for models of behavior. They lived in common with animals, a relationship reflected in their carvings. The killer whale was as close to them as an uncle.

The fifty supernatural killer whales in the waters around the Queen Charlotte Islands lived in houses arranged in rows, like the Haida, and some killer whale chiefs were more powerful than others, depending upon the prominence of the cliff or reef under which they resided. The more powerful killer whale chiefs had multiple fins reflecting their status. A model totem pole with a double-finned killer whale, carved by Charles Edenshaw, is in a glass case on display in New York, next to his model of the "Myth House." Supernatural marriage matched the Killer Whale Chief of each important reef with the Creek Woman of each adjacent stream that had an important salmon run. The fish—children of the union—were thought to leave their father's house and migrate, like a village of people, to the house of their mother, which was at the head of the stream.

Occasionally, one of the Haida would meet one of the killer whale chiefs while out at sea and he would go and visit his house. There were three cosmic zones—the sky world, the earth, and the underworld—and time was different in the cosmic zone of the killer whales. Thus, such a visit would feel as if it lasted four days, but on returning the Haida would find that four years had elapsed. I went south to see if I could arrange such a visit.

"How should you start a drawing then? Look at the subject as if you had never seen it before." That was the advice in one of the first drawing books I ever read, *The Joy of Drawing,* by Gerhard Gollwitzer. I thought it was good advice, about drawing and almost everything else, and it came to mind as I was waiting in Sandspit (population six hundred) for a flight south to the South Moresby National Park Reserve, the southern one third of the Queen Charlottes. There, I met Marty, a blonde, blue-eyed five-year-old with a huge smile, the son of Al and Marion Shaffer. Al would be my guide in the south. The afternoon before departure, Marty and I sat on a back porch and drew all the ani-

mals in the Queen Charlotte Islands. The Bear: The Charlottes have the world's largest. The Mouse: There are two
species of deer mice. The Eagle: The highest eagle nesting
densities in the world, except for Alaska. The Crocodile.
Okay, okay, that one might come as a surprise and it would
surprise a lot of biologists, Canadian Galapagos or not, but
Marty insisted and I was inclined to believe him. We also
drew the Mosquito, which looked terrifying and for good reason. Marty had decided he no longer wanted to go camping,
because every time he tried to poop in the woods the mos-
quitoes attacked his bum.

We had a three-step approach. He would start the drawing, I would finish it, and we would color it. We weren't
bound by any Renaissance conventions. There was no perspective and little detail, but there was always one character-
istic that made the animal unmistakably what it was, much
like native Northwest Coast art. The bear was as big as a
mountain. The mouse as small as a blade of grass. The eagle's
wings covered the sky. The mosquito had an enormous pair
of fangs, which, for Marty, expressed its spirit.

We saved the killer whale for last. Marty began by drawing
a dorsal fin, which reached the clouds. I added the body,
wide tail flukes, and had the whale leaping out of the water.

"What color should it be," I asked him. I was thinking of a
rich dramatic black and an ivory underside, but he had already started to color it.

"Purple," he said quickly.

Perfect, I thought.

Sandspit exists because Fletcher
Challenge, last seen approaching Paul Spong's Orcalab, has
been in the Charlottes since 1922. As described in one of

their colorful brochures, which included photographs of hik-
ers around a campfire, fishermen, and an orange helicopter
hauling timber, they intend to "continue contributing to the
area's economic stability in years to come." No wonder. A
single Sitka spruce log, thirty-six feet long and clear grained,
with eight growth rings per half inch, often used for guitar
and violin tops and now only a memory in Japan and Europe,
is worth as much as seventy-five thousand dollars. The
Fletcher Challenge Sandspit Division employs 110 people,
controls about 86,500 acres of forest, and over the next
twenty-five years plans to cut enough old-growth timber
each year to build 4,500 wood frame houses.

We left Sandspit behind and flew toward South Moresby,
which was declared a national park reserve in 1987, in a red,
white, and blue Grumman Goose, built in 1939 and looking
its age, but still sturdy as a Clydesdale. We felt every minor
gust of wind, every updraft and downdraft, and set a record
for slowest speed in the air without falling. The pilot kept the
Goose at fifteen hundred feet, until he suddenly soared to
scrape over the three-thousand-foot Cristobal Range, which
runs like a spine through the South Moresby Islands, and we
looked out on more than a thousand miles of convoluted
coastline, bays, inlets, lakes, and islands. The Pacific pounded
the rocks on the west coast. To the east, the bays were calmer
and the range of mountains sloped toward the shore, drop-
ping quickly from an alpine highland of low grasses, to rich
green forests of Sitka spruce and red cedar, ending in bog-
lands. In less than an hour, we touched down in Louscoone
Inlet, a fjord on the southwest coast. It was late afternoon.
The Goose landed in the water with the grace of a steam cal-
liope. The pilot drove it up on the pebble beach, where gear
for eight was dumped, then gouged out sand and gravel like a

hedgehog as he backed away from the beach and departed. The Goose moved so slowly on the inlet, flight did not seem to be one of the options, but, with a huff, a puff, it inched its way into the air, skimmed the spruce trees surrounding the inlet, and disappeared. Within minutes, the inlet had calmed and was glassy smooth. The air was sweet with spruce. I looked around for purple killer whales. There were none, but if there were killer whales to be seen, they might be out in Houston Stewart Channel. Anthony Island, which now has the Haida name Skungwai—Red Cod Island—sits at the channel's western end, on the edge of the Pacific Ocean.

The Haida traveled in great canoes, made from a single, clean-grained red cedar, in various sizes and proportions, some as long as seventy feet, seven or eight feet across, which could carry up to five tons of cargo or as many as three dozen people. I was not in one. I was in a polyethylene Nimbus "Puffin," a marine blue single kayak about sixteen feet long and twenty-seven inches across, built for stability, not for raiding and trading. My gear and food— including my dog tent—were stored in waterproof bags in two watertight compartments, fore and aft. A chart was in a waterproof case directly in front of me. The previous day we had paddled six hours in warm, sparkling sun, little wind and gentle swells, to a new campsite on a protected pebble beach, which put us just a few hours paddle from Ninstints. We had skirted the rocky shore of Louscoone Inlet and its south end, several times pushing through beds of thick, rubbery, brown bull kelp, which can grow up to a foot a day and is also called "elephant sperm," because of its spermlike shape. As we skirted the cliffs, I saw hundreds of dun-colored starfish, which had attached themselves in clumps to the jagged rocks just above the waterline. A few feet below, in crystalline water, were hundreds of bright, brick-red sea urchins, small balls of sharp spines. Sea otters once ate the urchins, along with crabs, fish, clams, and mussels, but the otter was trapped to extinction by early fur-traders and the urchins then flourished. Urchins eat practically any living bit of material that washes against their spines. Thus, they caused a decline in kelp beds, because a high percentage of the young replacement algae were eaten before they could grow to large kelps.

We had pushed off from our new campsite as the sun was rising, in order to reach Red Cod Island and the village of Ninstints by midmorning. To the west lay the Pacific; to the east, Houston Channel. We paddled easily. It was mild, with light clouds, little wind, and a faded sun. Al—the guide—was in an orange Puffin, lurking off to the side, hunched down with an implacable and powerful paddling stroke. He wore a turquoise baseball cap, which in the week I was with him he never once removed, and a barometer on his wrist. He used gray duct tape to cover the blisters on his feet. The others in the group—two hospital administrators from Calgary, three schoolteachers, and a publicist from the Calgary Zoo—were in three white and black double kayaks. It was my first extended kayak trip, but it took only a few minutes of travel before realizing that on wide open water, it was more a rhythm than a trip—a steady stroke, stroke—a rhythm that demanded a certain patience and humility. The kayak is close to the water and its small size makes it rather endearing. After a few days, it would feel almost like a piece of clothing, albeit usually a wet piece. Along the way, the sea was always changing and each stroke could bring something unexpected. Yet, if small was endearing, it could also seem like a cage. Once in, there was no getting out until the next stop, and the only way to get there was to keep pumping, with blistered hands, a wet butt, and wet feet.

As we approached Ninstints, I paused a few hundred yards away. The water had barely a ripple and was deep green from the reflection of the island's silhouette, which was covered in Sitka spruce. The shore was rocky brown and black and there was a narrow cut in the rock face, which led to an open cove and a pebble beach. As I paused, the others passed through the cut. From where I sat, I could see a few of the still-standing totems, gray sticks among the spruce, one of them listing to the right, as if ready to topple. I bobbed easily in the water. The poles reminded me of New York City. I wondered if one day, hundreds of years from now, someone wouldn't sit offshore in New York Harbor and look at the skyscrapers, listing against each other, gray from a distance. How is it, they'll wonder, how is it that a once thriving population could be reduced to nothing?

Five black and white pigeon gilmots settled into the water a few feet to my right. They, too, bobbed in the water, as if they were watching the island, five of nearly a million pairs of seabirds that breed on the islands, including storm petrels, cormorants, and tufted and horned puffins. Those seabirds lived on the ocean eating zooplankton, crustaceans, and fish, except during the summer nesting period, and all seabirds nested on the seaward side of the islands. Those five probably had their nesting sites on the opposite side of the island, which faced the Pacific. I read once that seabirds arrived "fully equipped through evolutionary change to fill practically every available niche offered by the land and sea." Fully equipped, however, did not mean they had an easy life. Changes in the temperature and direction of the surface waters, for example, could create a rapid turnover of plankton, leaving only a small bit available to seabirds. Mass starvation could follow. Other changes could bring on a "red tide," a rapid bloom of tiny, marine algae called dinoflagellates, toxic to humans and dangerous to seabirds. Heavy summer rainstorms could wipe out an entire year's chicks. If that wasn't enough, they also had to contend with the peregrine falcon, the majestic raptor that, in the South Moresby area, has the highest breeding densities for falcons of anywhere in the world. The falcon thrives in the cool, moist climate, nests in the immediate vicinity of a seabird colony and, traveling at some two hundred miles per hour, takes small- and medium-sized birds—such as those sitting in the water next to me—in mid-flight. One biologist has described the falcons as, "perhaps the most highly specialized and superlatively well-developed flying organism on our planet today." From the seabirds' point of view, the falcon is a "whoosh," a violent and sickening thump, and death. Imagine being hit by a high-speed train. That is the evolutionary niche the pigeon gilmots occupy.

The Haida, like many Native Americans, and even killer whales, for that matter, have also been described as being perfectly adapted and attuned to their environment. Like those on Red Cod Island, with their Chief Ninstints—which means "He Who Is Worth Two"—they lived in isolated villages of a few hundred people. In Ninstints, there had been about four hundred of them along a narrow stretch of beach and rock-bound cove. On one side, they faced treacherous ocean currents. On the other, dense forest. Canadian anthropologist Wilson Duff has described this world as being "sharp as a knife," cutting between the depths of the sea, which to them symbolized the underworld, and the forested mountainsides, which marked the transition to the upper world. They embellished the knife-edge with carved monuments and brightly painted emblems, with creatures of the upper and lower worlds, a balanced statement of the forces of the universe. The crucial thing was to maintain harmony between the cosmic zones, because disharmony would be reflected in an unfortunate event: the failure of a fish run, a severe storm, or the arrival of pestilence. I like to think that they were adapted and attuned enough to know their position was precarious, which is perhaps the greater wisdom. They realized that, even if other native groups might look upon them as fierce warriors and great craftsmen, they were more like seabirds, tossed by storms, set upon by raptors.

The pigeon gilmots suddenly departed, all five running across the water before taking flight. I paddled toward the island and rode the incoming tide through the cut in the rocks, popped out the other side, and glided easily toward the beach. Red Cod Island is tiny, just 350 acres. On the western side, the winds can be some of the strongest in Canada, and the Pacific nervously pounds the rocks, but the east-facing cove was gentle and well-protected. Above the tide line, the weathered totems watched the cove like brooding sentinels. Seven poles were removed from Ninstints in the late 1950s to the University of British Columbia Museum of Anthropology, a project in which Bill Reid participated. Today, some sixteen poles remain. Wind and rain, grass and spruce continue to sculpt the soft cedar. They have taken on new shapes—gray, crumbling, and punched through with new growth. I spotted the killer whale totem to my right front and paddled hard toward it, running the kayak up on the beach. It was a mortuary pole. At one time, there had been a box atop the pole, with human remains. The pole stood straight, but it was stumpy and rotted in places. The killer whale was the only recognizable figure. Its flukes were folded down. A bulbous protrusion emerged from its blow-hole, which looked to me like a frog. The midmorning sun turned the grass around the pole a golden green.

We stowed our gear. A few minutes later, Dick Belles arrived. He was one of the Haida "Watchmen," who was staying on the island for two months. He was a husky man and wore burnt orange "Husqvarna Chain Saws" suspenders over a black and white plaid shirt, jeans, and heavy work boots. The Parks Department had wanted the Watchmen to wear uniforms, rather than chain saw suspenders, but Dick said the Haida had refused. "This belongs to us, not to them," he said. He also said that he had arrived on the island for the first time in his life the day before. He held up for our examination a small, silver killer whale medal, which hung from a chain around his neck. His loyalty was divided between the lumber company, his employer for the past thirty-two years, and his clan, the killer whale. He said that someone who lived a normal life was entitled to one dorsal fin; a half-decent life was entitled to display two. So, his medal said, symbolically, that he was from Masset, of the killer whale clan and "a half-decent man."

We followed him as he began to describe the poles and the story associated with each one. I was eager to hear the story of the killer whale, but we first passed a grizzly bear mortuary pole, identified by its large snout, and a black bear, badly decayed. We passed an eagle mortuary pole, then another bear pole with three watchmen encircling the top, followed by a mortuary pole worn completely smooth and another black bear pole. Dick referred continually to a paperback book written by George F. MacDonald entitled, *Ninstints, Haida*

World Heritage Site, a condensed version of a much longer book entitled *Haida Monumental Art*, which referred to Ninstints as one of Canada's "holy places" and described the efforts made by the Canadian government to preserve the site. As a World Heritage Site, it is on the same list as the Notre Dame Cathedral, Angor Wat in Cambodia, and the Mayan cities in the Yucatan Peninsula, although Dick, our guide, told us that the Haida weren't interested in preservation and insisted that the poles "will die a natural death."

We finally stood before the killer whale mortuary pole. My kayak was just below us on the beach. I had seen a photograph of the pole, taken in 1901 by Charles Newcombe when he was scouring the islands for Franz Boas, and that photograph had shown a human figure tucked between the downturned flukes. The human face had long since been sculpted away by wind and rain. I thought Dick might supply some short parable about the pole, about the fading of the human face. Instead, the Haida Watchman described the killer whale pole as a grizzly bear. As he did, though, two fat and rumpled ravens, sitting in a nearby spruce, suddenly started an antic dance, hopping up and down and croaking to one another. It wasn't the Watchman's mistake. The ravens had done it. They had tricked us and it was clear they had enjoyed it immensely.

About half an hour later, Dick was interrupted by a film crew from the Canadian Broadcasting Company, who were making a public service film for Tourism Canada. We had seen their gray zodiac boat on the beach. Dick and the director of the film had met the week before, when the director was doing a safety film for Dick's employer, MacBlo. Now, the director wanted to do "the

Watchman scene." He wanted Dick to talk, talk about anything, while he filmed. The sound would be dubbed over
later. So, Dick started to talk to us about how the forestry
practices of MacBlo were continually misrepresented to the
general public. "We know what's good for this island and
what's not and it's not what my friend David Suzuki thinks,"
Dick said. David Suzuki is the best known environmentalist
in Canada, whose weekly television show, *The Nature of
Things*, reaches nearly a million viewers. He also narrated a
1987 film on South Moresby entitled *Wilderness Under Siege*,
which was entered into the parliamentary record by the federal environment minister and helped generate public support for preserving the wilderness. As Dick talked about his
employer, MacBlo, he was filmed as the Haida Watchman, as
if he were discussing his killer whale crest.

When the scene was completed, Dick wandered away. I
started to poke around, but as I began to follow a path that
led to the west side of the island, I was stopped by the fran-
tically waving film crew. Did they want me for something?
They did. I was about to walk through their next scene. I
stopped. A young Haida woman, with a solemn face, olive
skin, and long, shiny black hair, was standing a few steps out
of camera range. After I'd been halted, she composed herself, stepped into the sunlight being reflected by two large
white panels, looked seriously into the camera, and said,
"The Haida have walked with care here for centuries and
would like you to do the same . . ." She stopped and shook
her head. Apparently, she had forgotten her lines. She
stepped away from the camera and quietly repeated her lines
to herself, moving her lips. The director, a young, bearded
man, in jeans and a red and white sweater, called out, "Once
more, Mary Ann." She composed herself for the second
time, then again stepped into the reflected light. "The

Haida have walked with care on this land for centuries and would like you to do the same," she said sincerely. "Please pack out your garbage."

"Their genius has produced monumental works of art on a par with the most original the world has ever known," wrote anthropologist Marius Barbeau, which is why the Canadian government wants to preserve the site. The Haida believe it should be left to decay. We Europeans want to capture and contain, to call a spot holy and fend off death, but the moment of creation is the holy moment. The site is only a reminder of how we—the Haida, all of us—struggle to carve out a space in which we can live, and all of those efforts, even the greatest of artistic works, will eventually be overgrown with creeping vines.

I sat in the warm sun and drew the pole. It had been described as a grizzly, but it was a killer whale, with a completely unrealistic long snout. I could feel it as I drew it: clean lines, bulging eyes, and ferocious teeth. It seemed to me just that it would be allowed to melt away, beautifully carved by the wind and rain. I looked closely into the cracks of the weathered pole, beneath the flukes, where the man's face had once been. The human spirit, which had once been inside the whale, had long since disappeared. In its place, I saw a tiny purple flower, a harebell, bursting with life and defiance.

Before I'd ever been in a kayak, I'd read the following two bits of advice: First, "Learning to recognize and avoid situations beyond your abilities is a primary skill in sea kayaking"; second, "What do you do if you are caught in a storm? The best tactic is usually to get ashore." I'd read both of those in John Dowd's *Sea Kayaking*, often referred to as the "kayaker's bible." I always remind myself of those two bits of advice whenever I plunk down into the cockpit of a kayak, and I did it when we left the protected cove of Ninstints and headed north. As we skirted its west coast, we easily rode five-foot Pacific swells and watched a barking, chocolate-brown Stellar sea lion haul himself out on the rocks. I considered that advice later, when the weather was transformed from brilliant sun to nasty storm in about the time I took for a deep breath. We were looking for killer whales—or I was, anyway—farther into Houston Stewart Channel, circling the craggy, volcanic Gordon Islands, fighting the wind, paddling past three tufted puffins, fat, globular seabirds with bright orange beaks and yellow feathers swept back behind orange rimmed eyes. They ignored us and appeared to be riding out the storm. I could have touched them with my paddle.

Al had guessed that the wind was going to shift to the east as the storm strengthened. If it had, it would have blown us home to our campsite. As we paddled around the island, through choppy two-foot waves broken by whitecaps, the high wind was in our face. We stopped twice in small clefts in the rocks, to rest out of the wind, although during the final stop the refuge was swamped by a swell that surged into the cleft and pushed the kayaks onto the rocks. We pushed each other and paddled wildly to finally time the surge and ride it out of the cleft, back into the wind and rain. Then, we began what should have been a half-hour crossing of open water to the campsite.

Al had guessed wrong, though. The wind never turned. We paddled straight into it, fierce and cold, a rain of needles on my face. I leaned into it, trying to keep a low profile and put my body into each stroke. I'd worn a pair of rubber gloves over some liners, but my hands were quickly stiff from cold. I could lift my eyes only high enough to see the sea just beyond the bow, gray and carved by whitecaps. The winds were rising. More and more whitecaps formed. The rule of thumb is, whitecaps form at about twenty miles per hour, but that wind had probably been much higher, since it was blowing in the same direction as the tide and, thus, it would take a higher speed to form the whitecaps. Higher gusts pushed the nose of my kayak to my left or south. Every fourth stroke, I used a wide sweeping stroke to keep myself on course, which became a metronome in my head—stroke, stroke, stroke, sweeeeeep—keeping my eyes on the gray sea just over the bow and no further, lest I see just how much farther we had to go.

The metronome in my head had been working for hours when the Gordon Islands finally faded to the rear. Buffeted by the waves, head down and squinting into the wind, I kept the nose of my kayak pointed at a small spit of land and stroked, confident that the kayak was as stable as the Queen Mary, but also knowing that if I stopped to think for even a moment, I would feel a dull ache in my shoulders from the steady stroke, stroke, stroke, sweeeeeep. When a sharp gust of wind blew my rain hood back with a snap, I ignored it. Seconds later, my hat blew off. I didn't look back, feeling that if I paused, I would never regain my rhythm, the nose of the kayak would be pushed off its track, and I would follow my hat. I thought of it as an offering to the Chief of the Underworld.

One by one, the other kayaks disappeared around the spit of land. Then, Al, with his implacable stroke, was also gone. I was alone in a gray sea, a painful rain in my face, long white tendrils of water scattered in every direction by the wind, trying to keep my rhythm and my eyes on the front of my kayak. The wind strengthened. I continued to stroke, continued to think of nothing but rhythm, but, as difficult as it was to tell in such a choppy sea, I slowly realized I was no longer moving forward. Each stroke was barely holding me in place. I continued paddling that way for perhaps a minute, not wanting to believe it, but not moving, and I thought to myself, "Well, now what?" I was thinking, rather clearly, that perhaps I was destined to follow my hat. Yet, as I sat suspended, neither moving forward nor backward, I felt something more. I felt tiny in the face of the storm, and as I began to imagine myself tossed, turned, smashed on the rocks like a breaker, I felt I was a breaker, as large as the storm itself, which was threatening to send my kayak tumbling toward the South China Sea. I felt I might be crushed on the rocks, but I was the rock itself. I had no strength, but I felt strong. I felt that I would never know, but I still had questions.

"Well, now what?" The wind diminished. Or, did I gain strength? My kayak jerked forward. I was stroking and moving. I crawled around the spit, staying clear of the rocks, where the water was swirling and looked as if it could swallow a cruise ship. On the other side of the spit, it was relatively calm, protected from the wind. Al was waiting for me, hunched down in his orange kayak, looking unperturbed. The other kayaks were already beached. I wondered, how did they get to the beach so quickly? The half hour crossing had become two hours, but how long had it taken me to get around that spit? As I stroked toward shore, Al paddled alongside. "I won't forget that for a while," he said.

Once ashore, I found that my tent—the dog tent—which I had pitched behind an enormous, smooth bone of a log on the beach, was covered with sandhoppers, small, reddish, shrimplike creatures that live in the kelp. Clumps of them had collected on the outside of the tent and the sand pulsated with thousands more. The sand was alive. I started to scrape the hoppers off, when I realized that the sandhoppers were actually retreating up the beach in a huge pulsing wave. My tent was an obstacle, not a destination. They were passing over it, headed for higher ground in the face of the storm and high tide. The foam of the breakers on the beach was brown, rather than white, from the churned-up bottom. A quick check of the tide tables and some conservative estimates of the possible storm surge, and it was clear my tent would be swamped in a few hours. So, after scrapping off clusters of sandhoppers, I followed their instinct to higher ground and joined them in the spruce forest, where I quickly pitched the tent, stripped, put on a pair of clean, dry, long underwear, and collapsed in my sleeping bag. I lay flat. I waited for warmth. I listened to the rain and wind and the waves thundering on the beach below. I felt as if time had stopped and the universe was contained in my tent. The Haida thought the vault of heaven was an enormous skin tent and explained the northern lights as "the skin of the sky is burning." My own vault of heaven stretched above me. Its skin snapped in the rain and wind. I was cold and I slept.

The following afternoon I crawled out of my universe and stood on the beach. The storm had spent itself. The tide was receding, although brownish, gritty waves still crashed on the beach. The slow-moving mass of sandhoppers were making their way back to the beach, except for those that had taken refuge inside my tent, who appeared to be quite content. A pile of four drift logs, white, gleaming, and two feet thick, had settled on the spot where my tent had once been pitched. The thin black forest silhouette on the horizon was socked in by mist and leaden clouds. I knew that, somewhere in that mist, a team of biologists was looking for the mystery whales. They hoped to eventually dispose of that phrase, but I ran it over my tongue a few times and I said it aloud: Mystery whales. I liked the sound of it. I sniffed the air and stuck out my tongue. It tasted delicious.

Manhattan is from the
Native American—Manahatta—and depending upon the dialect and the source one consults, could mean "cluster of islands with channels everywhere" or "the island where we all became intoxicated." Both have some truth in them. Walt Whitman said it meant, "the place encircled by many swift tides and sparkling waters." When I returned, I wanted to see the city from those waters in a kayak, but it took several attempts. The first time, my girlfriend Karen and I planned to

go with a kayak outfitter from upstate New York, but the night before our trip I was awakened six times by nightmares of Karen drowning. I remember two. In the first, her kayak had capsized and she was upside down in the water, unable to get out. In the final one, a tugboat was bearing down on us and I said, confident and knowing, "Don't worry, he sees us." Then, the tug ran us down. I was flailing around in the water and I awoke, rolled over in bed, and said to Karen, "I don't think we should go, today." We were set to try again a few weeks later, but a thunderstorm swept through the city, and I will never again be caught—intentionally—in a kayak during a storm. We postponed, again.

I sometimes find that a casual idea stays casual only until it has been thwarted. I quickly found that it was more difficult to rent a kayak in New York City than to park a car. A close friend intervened with news of a kayak I could borrow, if I was willing to haul it. There were just a few drawbacks. It had been sitting unused in a backyard for fifteen years, so it might not float, and it had no seat. Still, it seemed worth a try. We substituted the passenger seat from an MG and I bought a new lifejacket, just to be certain that one piece of equipment was seaworthy. We lashed it to his truck and drove to Liberty State Park, New Jersey. The plan was, I would paddle around the harbor, he and Karen would go off to the Ellis Island Museum, then together we'd help him move a couch to an apartment in Queens. Unfortunately, by the time we reached the park in midafternoon, the Indian summer weather had turned to fog and a hard rain. We walked around in it. We cursed it. We sat and warmed ourselves with hot chocolate and laughed and said, Can you *believe* this?

We waited for perhaps an hour. Then, without a word, we both rose, walked back to the truck, and unlashed the kayak. We hauled it over a railroad siding, which must have at one time been used for cargo, and across sharp, oil-blackened rocks covered with yellow-green algae. I pushed off, quickly paddled clear of the boat landing, then stroked along beside the sea-wall of the park until I was into the harbor. The water was calm, but the paddling wasn't easy. The wooden paddle, which had been sitting out in the back yard for as long as the kayak, had broken. Immediately prior to launch, I sawed off a one-foot length of wood from a canoe paddle and duct-taped it to the kayak paddle as a brace. It worked, but it was like paddling with a railroad tie.

From the harbor, New York was a city of islands surrounded by swirling waters, with 578 miles of shoreline. The enormous Hudson was to my left, a full mile wide where it washed the west side of the city; Manhattan Island, Staten Island, Long Island, Liberty, Ellis were all in front of me, and if I went farther and turned left, I'd see Randall's, holding up a bridge, and the city jail on Rikers. Blue and yellow ferries were probing the fog, headed for Staten Island. I pushed on toward Liberty Island and the statue by Bartholdi with its uplifted torch. A green and black tug churned across my bow, pushing a white Lehigh Portland Cement Company barge. I easily rode up and over the wake.

In John Kieran's *A Natural History of New York City*, I read that a young sperm whale once followed a steamer into New York Harbor and died when it became stranded in Brooklyn's Gowanus Canal. I followed that story into the archives of the New York Public Library to flesh out the details, but found neither data nor myth. Similarly, I looked for but failed to see whales that day in the harbor, and settled instead for dozens of herring gulls, two feet long with four- or five-foot wingspans, which hang around sewer outlets and descend by the thousands on city dumps, like the nearby Fresh Kills landfill, just to the south, which may well be one of the highest points of land on the East Coast south of Mount Desert Island, Maine, by the end of the century. Yet, even in their thousands, they don't outnumber the pigeons, which arrived with the early colonists and thrived, like so many immigrants to the city, even though, like any other city resident, they must be alert for predators. The peregrine falcon, also found in the Queen Charlotte Islands, has returned to the city after being absent for nearly thirty years, with some ten pairs nesting in high buildings and bridges. The pigeons, as one biologist described it, provide a "rich base of prey." That's life for pigeons and pigeon gilmots, in the city or in the isolation of the Queen Charlotte Islands. They think they've found their niche and suddenly they're lunch.

After I'd returned to New York, I read that in Vancouver, killer whales had again made headlines. A calf was born to Bjossa, described as a "perfect birth" and named K'yosha. Sadly, it did not have a perfect life and died after just ninety-seven days, spending its last hours stricken by an abscess near its brain and banging its nose on the concrete of the Killer Whale Habitat, once hard enough to shatter its jaw. The famous Hyak did not leave an heir.

I also had a letter from Jim Hart, the Haida carver. I had written to him from Victoria, British Columbia, where I saw a watercolor painting that I considered buying, even though for me it was expensive—one hundred dollars. After deciding not to buy it, I had immediately written to Jim and explained that I had finally arrived at a price. If I was willing to consider paying a hundred dollars for a watercolor, I reasoned, I was willing to spend the same amount for his drawing. If he thought that was fair, he should let me know. In his letter, he referred to my offer as an "opening bid," which made me laugh out loud. The frog sculpture would eventually be one of the "Haida statements," he wrote, so he wasn't going to accept my bid immediately. "Remember that raggy ol' drawing is going to be a valuable chunk someday," he wrote, driving the price even higher. He said he intended to one day come to New York to see the carvings in the Natural History Museum and added that he would soon be training some young people in totem pole carving. "We'll make all anthropologists happy, all archaeologists of the future happy," he wrote. "What a life."

I wrote back with some New York advice: Frame the drawing, offer it to the realtors who commissioned the sculpture as a "start-to-finish package—the Haida artistic mind at work," and then ask them for the moon. I haven't heard from him since, so I assume the negotiations are ended and the realtors outbid me by an extraordinary amount. I hope they did, anyway. I wonder if they know its real value, which can't be calculated, a statement giving order to the world where otherwise there might be none, a carving that can force one to ask, as it did me, "Where does one creature end and the other begin?"—a question that can easily be as perilous as a kayak trip on a stormy sea.

I came in from the harbor in late afternoon, paddling through a thin rain. The sky was a thousand boiling shades of gray, and from my spot in the harbor Manhattan was wrapped tightly in fog and rain clouds, a package without lights or detail, the buildings black silhouettes, as if formed of Chinese ink spread on fine rice paper. Find a slightly different angle and the city can be seen again for the first time. That day, the city was as beautiful and mysterious as I have ever seen it.